CHEMISTRY OF GASEOUS STATE

CHEMISTRY OF GASEOUS STATE

By
M.S. YADAV

ANMOL PUBLICATIONS PVT. LTD.
NEW DELHI - 110 002 (INDIA)

ANMOL PUBLICATIONS PVT. LTD.
4374/4B, Ansari Road, Daryaganj
New Delhi - 110 002
www.anmolbooks.com

Chemistry of Gaseous State

First Edition, 2001

ISBN 81-261-0999-8

PRINTED IN INDIA

Published by J.L. Kumar for Anmol Publications Pvt. Ltd., New Delhi - 110 002 and Printed at Tarun Offset Press, Delhi.

Contents

Preface

This book "Chemistry of Gaseous State" is a book meant for students of B.Sc. (Hons) and M.Sc. students. The book covers most of the aspects of gaseous behaviour and the various topics such as gas laws, kinetic theory of gases etc. included in the syllabus for degree classes of Indian University. The book contains 8 chapters and each topic has been discussed in a simple and easy to understand way.

The book will be of immense value to students of chemistry, physics and applied sciences. The book is also likely to be very useful to the students of engineering colleges. In preparing the book I have consulted the works of many authors and am grateful to all of them.

I would like to express my thanks to Mr. J.L. Kumar, Managing Director, Anmol Publications Pvt. Ltd., New Delhi for bringing out the book in such a good form in such a small time.

Author

1

Gaseous State and Physical Properties of Gases

1.1 Introduction

The gaseous state is one of the three states of matter (i.e. solid, liquid, gas). Water can exist in either a solid (ice), a liquid (water) or a gaseous (steam) state depending as the temperature and pressure. Depending on the temperature and pressure a substance can exist in any of these three states. In this volume we are mainly concerned with the study of the behaviour of substances when they are in gaseous state. To begin with we recapitulate that in case of gases the particles are completely free to move about and so the *gases have neither a definite shape nor a definite volume* and thus a gas is defined as *that homogeneous state of matter which has neither a definite shape nor a definite volume and takes up the shape and volume of the containing vessel.*

The three states of matter (viz. solid, liquid and gas) have different characteristics. The essential difference between three states of matter is in terms of the energy contents.

Some of the important characteristics of gases are:—

(i) *Homogeneity*. A gas is homogeneous i.e. its composition in uniform throughout.

(ii) A gas has neither a definite shape nor a definite volume.

(iii) The gas molecules exert pressure equally on the walls of its container. The pressure increases with the rise of temperature.

(iv) Gases can penetrate easily into one another, forming again a uniform mixture. This process is known as *diffusion*.

(v) It is possible for a gas to expand to any extent and it remains homogeneous even during expansion.

(vi) The densities of gases in low.

(vii) Under increased pressure, a gas can be compressed to a smaller and smaller volume until it liquifies.

The study of gases provides an ideal introduction to physical chemistry. The two principal components of physical chemistry are clearly identified. One is the organisation of the experimental results obtained from the studies of the "ordinary" or macroscopic world. The other component is the study of the atomic molecular world that underlies this macroscopic world. Physical chemistry deals with the study of these aspects and the bringing of them together to give a molecular level interpretation of the observed macroscopic phenomenon.

The behaviour of gases is often quite simple and many physical properties are found to be quite identical for all the gases and so it is possible to state some simple rules for the behaviour of gases. The gaseous state is the simplest of the three states of matter, to analyse in terms of an atomic model and is essentially completely understand today.

In the following pages an attempt will be made to study the pressure volume temperature behaviour of gases. These studies will provide a base from which many practical and penetrating studies stem. The practical aspects will be apparent when we go from behaviour of *ideal* gases to those of *real* gases. In thermodynamics we make much use of these simple ideal gas relations.

1.2 Gases, their Behaviour, and an Explanation of their Behaviour

In the following table are listed some gases:

Name	*Formula*	*Colour*	*Odour*
Ammonia	NH_3	Colourless	Penetrating
Carbon dioxide	CO_2	Colourless	Odourless
Carbon monoxide	CO	Colourless	Odourless
Chlorine	Cl_2	Pale yellow	Irritating
Helium	He	Colourless	Odourless
Methane	CH_4	Colourless	Odourless
Neon	Ne	Colourless	Odourless
Nitrogen	N_2	Colourless	Odourless
Nitrogen dioxide	NO_2	Red Brown	Irritating
Oxygen	O_2	Colourless	Odourless

It can be easily seen from the above table that most of the gases are colourless but chlorine and nitrogen dioxide are coloured. Nitrogen dioxide lends its brownish yellow colour to smoggy air. Most gases are colourless and therefore invisible. Nevertheless, gases are enormously important. We must breathe oxygen to survive. Gases make up the air of the atmosphere, and without them this planet would be a very dead place, like the moon.

1.3 The Nature of Gases

The following properties are common to all gases:—

1. *Gases fill a container completely and uniformly.* When you inflate a tyre all parts of the tyre are uniformly filled. In contrast, a liquid fills only the bottom of a container.

2. *Gases are compressible*
3. *Gases have low densities.* The density of a typical gas is 1g/L, whereas the density of a typical solid may be 2000g/L. Thus a solid is usually 2000 times more dense than a gas. A typical liquid may be 800 times more dense than a gas.
4. *Gases exert a uniform pressure on all inner surfaces of a container because they completely fill their containers.* Solids and liquids exert a pressure on as much of the container as they touch.

1.4 Gas Pressure

Though we all are familiar with pressure yet we cannot define it accurately. For example, when you press a finger against your nose, you might say you feel the "pressure" from your finger. The concept of atmospheric pressure is not very different. The atmosphere around our earth contains an estimated 5×10^{18} kg of gaseous substances. This mass, exerts a force on every part of the earth's surface because of the earth's gravitational attraction. The atmosphere pushes on the earth like your finger pushes on your nose.

Pressure is defined as force per unit area. The pressure exerted by the atmosphere on the earth is *atmospheric pressure.* The atmospheric pressure is greatest near the earth and decreases with increasing altitude.

The device generally used to measure atmospheric pressure is *Barometer*.

A mercury barometer (Fig 1.1) is constructed by filling a glass tube (about 1 metre long) with mercury, a metal that is a liquid under ordinary conditions. The tube is then inverted in a dish of mercury. Some of the mercury drains from the tube into the dish but considerable portion of it remains in the tube. A *vacuum* occurs above the surface of mercury in the tube because some of the mercury has left the tube and air has not been able to replace it. At sea level the average height of the mercury in the tube is 760 mm, but the height varies at different levels because it is a direct measure of atmospheric pressure. This pressure is exerted downwards on the surface of the liquid. It is also transmitted through the liquid and exerts a pressure upwards at the rate of the column of mercury. The upward pressure supports the column. If the pressure of the atmosphere increases or decreases, the column's height will increase or decrease correspondingly.

The most frequently used units of pressure are based on the height of the mercury in such a column. The unit **(mm of Hg)** *millimeters of mercury* is also called **torr** (after Evangelista Torricelli, who invented barometer in 1643). It is a unit of pressure equal to that exerted by a column of mercury that is exactly, 1 millimeter high, one *standard atmosphere* **(atm)** is another unit of pressure equal to exactly 760 mm Hg

1 atm = 1.01325×10^5 **pascal (Pa)**

Pascal (Pa) is the 5.1 unit of pressure

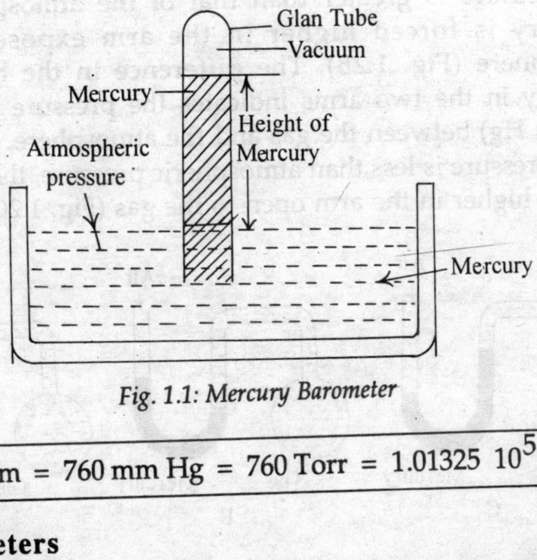

Fig. 1.1: Mercury Barometer

1 atm = 760 mm Hg = 760 Torr = $1.01325\ 10^5\ P_a$

Manometers

All gases exert a pressure on any surface that they contact. For example, the gas inside a child's balloon exerts a pressure and pushes the walls of the balloon outwards. If it is assumed that the balloon is perfectly elastic, the outward movement will only stop when the gas pressure within the balloon is equal to the atmospheric pressure. Thus if you have measured the atmospheric pressure with the help of a barometer, you have also determined the gas pressure within the balloon.

A gas enclosed, in a non-elastic glass or metal container also exerts a pressure, but to measure it we will need some other device. In the laboratory we use a **manometer** to measure the pressure of a gas in such a container.

A manometer consists of an open ended U-tube containing mercury (Fig. 1.2). The difference in the height of mercury in the two arms of the U-tube relates the gas pressure in the container to the atmospheric pressure. If the pressure of the gas is identical to the atmospheric pressure, the mercury level in the two arms of the U-tube are equal because the gas and the atmosphere are pushing on the mercury with the same amount of force (Fig. 1.2A). If the gas pressure is greater than that of the atmosphere, the mercury is forced higher in the arm exposed to the atmosphere (Fig. 1.2B). The difference in the heights of mercury in the two arms indicates the pressure difference (in mm Hg) between the gas and the atmosphere. However, if gas pressure is less than atmospheric pressure, the mercury level is higher in the arm open to the gas (Fig. 1.2C).

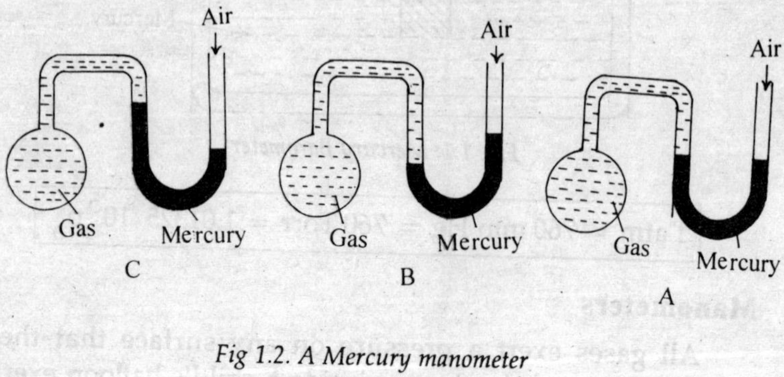

Fig 1.2. A Mercury manometer

1.5 THE GAS LAWS

Around the middle of seventeenth century, several scientists explored the quantitative relationships between a gas pressure, volume, temperature and amount. For example, they were able to predict the volume that a certain gas would occupy at a specified temp. and pressure. It was found that all the gases, *regardless of chemical composition*, obeyed these mathematical relationships as long as the low temperatures and high pressures were avoided. Here we will take up some of these gas laws because these gas laws led to the formulation of the kinetic molecular theory.

BOYLE'S LAW (PRESSURE AND VOLUME)

The English Chemist Robert Boyle (1627-1691) experimented with the compressibility of air and discovered the law that is now named after him.

According of Boyle's law, "*Temperature remaining constant, the volume of a given mass of a gas is inversely proportional to its pressure.*"

This law was given by Boyle in 1662.

It can also be stated as '*The volume of fixed molar amount of a gas at a given temperature is inversely proportional to the applied pressure.*" The term *inversely proportional* means that one of these quantities will double when the other is h alved. The mathematical equation for this law is

$$\text{Pressure} \times \text{Volume} = \text{Constant}$$

$$PV = a$$

Where a is a constant for given temperature and given amount of gas.

To appreciate this law, consider Figure 1.3. It will be assumed that the temperature and the amount of gas do not change during the experiment shown in the figure. As we begin the pressure is 1.0 atm. and the volume is 4.0 L. When the pressure is doubled by increasing it to 2 atmospheres, the volume is halved i.e. it becomes 2.0 L. If the pressure is doubled again, the volume is halved once again. This demonstrates the inverse proportionality between pressure and volume of a gas.

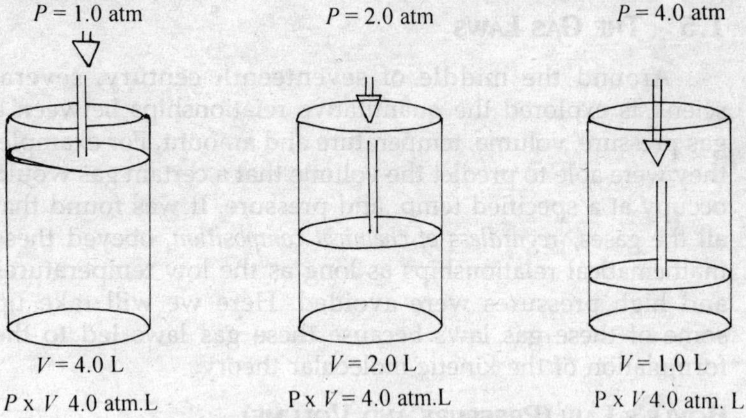

Fig. 1.3. The effect of pressure on the volume of gas. Doubling the pressure on a gas enclosed in a cylinder with a moveable piston cuts the volume of the gas in half.

It is seen from the relation $P \alpha \frac{1}{V}$ that the pressure of the gas is not linearly related to its volume at a constant temperature. However, since $P \alpha \frac{1}{V}$, we can write p = constant$(K)/V$

or $P = K \times x$ (where $x = \frac{1}{V}$). The relation $P = K.x$ is an equation of a straight line, so a graph of P and $\frac{1}{V}$ is a straight line (Fig. 1.4)

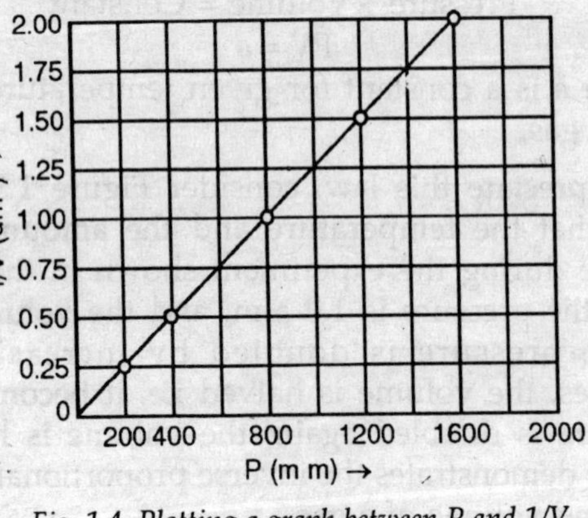

Fig. 1.4. Plotting a graph between P and 1/V.

Similarly the plot of PV versus P or PV versus V will also be a straight line. This is illustrated by the following examples (Fig. 1.5.) :

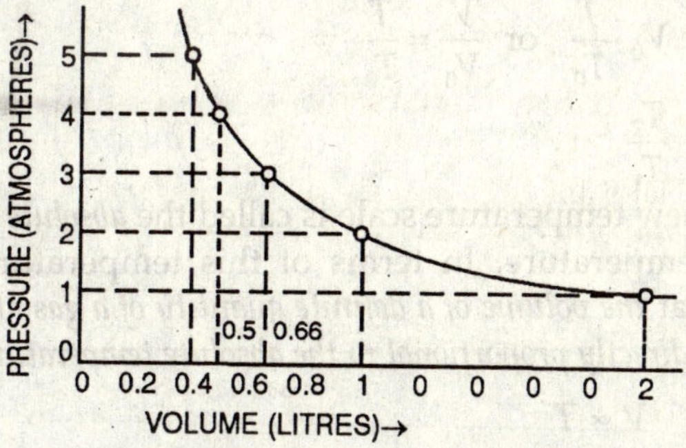

Fig. 1.5. Plot of volume versus pressure of a gas at a fixed temperature.

Charles Law (Volume and Temperature)

After Boyle completed his studies, chemists and other scientists continued to examine gases and their properties. Jacques Charles (1746–1823) was the first to inflate a balloon with hydrogen and make an ascent with it.

Charles found that the volume of a fixed amount of gas increased when he increased its temperature, as long as the pressure remained constant. If he cooled the gas under same conditions, the volume decreased.

The influence of variation of temperature on the volume of a gas was investigated by Charles (1787) and also by Gay-Lussac (1802) independently.

Charle's Law or *Gay-Lussac's Law* states that, "*at a constant pressure, the volume of a gas increases by 1/273 of its volume at 0°C for every degree centigrade rise in temperature.*"

If a given mass of a gas has a volume V_o at 0°C, then the volume at t°C and the same pressure will be given by the equation

$$V_t = V_0 + t.\frac{V_0}{273} = V_0 \frac{(t+273)}{273}$$

We may now define a new temperature scale such that any temperature T on it will be given by $T = 273 + t$ and 0°C by $T_o = 273$. Then the above equation becomes

$$V_t = V_0 \cdot \frac{T}{T_0} \quad \text{or} \quad \frac{V}{V_0} = \frac{T}{T_0}$$

$$\text{or} \quad \frac{V_2}{V_1} = \frac{T_2}{T_1}$$

This new temperature scale is called the *absolute* or *Kelvin* scale of temperature. In terms of this temperatures scale, we find that *the volume of a definite quantity of a gas at constant pressure is directly proportional to the absolute temperature.* Hence

$$V \propto T$$

$$\text{or,} \quad V = \text{constant} \times T$$

When volumes are plotted against absolute temperatures (at constant pressure), straight lines, called **isobars,** are obtained which on extrapolation converge to zero volume. (Fig. 1.6).

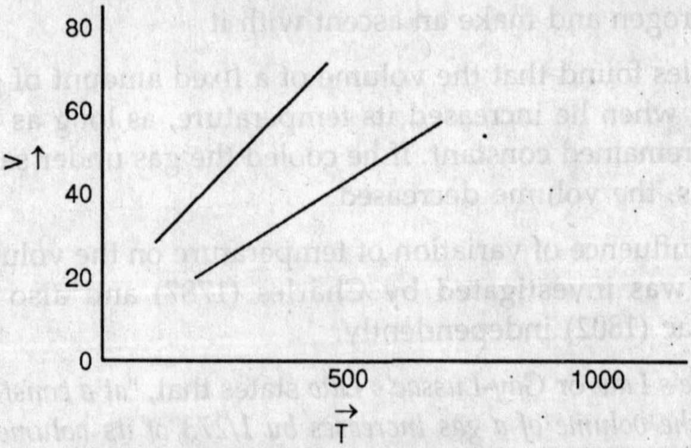

Fig. 1.6. Plot of volume against temperature Isobar ($P_1 > P_2$).

Modern chemists prefer the use of absolute or Kelvin scale of temperature. –273.1.5°C is taken as zero of the Kelvin scale. The temperature T read on this scale is written as T° Kelvin or *TK*.

Kelvin temperature is related to the centigrade scale as under :

Kelvin temperature (*K*) = Centigrade temperature (°C) + 273

Thus 20°C = 20 + 273

= 293 *K*

Charles law can be represented by the equation

Volume = Constant x Kelvin temperature

If we write T for the temperature on Kelvin scale our equation becomes

$$V = b\,T$$

If you are considering a sample of gas whose volume and temperature and changing then we can write

$$\frac{V_1}{T_1} = b = \frac{V_2}{T_2}$$

or $$\frac{V_1}{T_1} = \frac{V_2}{T_2}$$

The equation in this form is used for problem solving.

Example 1. *A given mass of a gas occupies 200 cm³ at 127°C. What volume will it occupy if its temperature is lowered to 27°C at constant pressure?*

Solution.

Here V_1 = 200 cm³ V_2 = ?

T_1 = 127+ 273 = 400 *K* T_2 = 27 + 273 = 300 *K*

Now $$V_2 = \frac{V_1 T_2}{T_1} = \frac{200 \times 300}{400} = \mathbf{150\ cm^3}.$$

If we note down the volume (*V*) of a certain mass of a gas at constant pressure at different temperature and draw a graph between *P* and *T* (absolute temperature), it will be found to be a straight line.

Avogadro's Law (Volume and Moles)

Boyle's law gives a relationship between the pressure of a gas and its volume (at constant temperature and amount) and Charles law gives a relationship between the volume of a gas and its temperature (at constant pressure and amount).

According to **Avogadro's Law,** "*equal volumes of any two gases at the same temperature and pressure certain equal moles of molecules*." This relationship is named for the Italian Chemist Amadeo Avogadro (1776-1856), who postulated it in 1811 during his studies of gas volumes.

This law can also be stated as, "*the volume of a gas is directly proportional to its amount in moles as long as the temperature and pressure are constant.*" Thus we can write

$$V = dn$$

where d is a constant and n is the number of moles.

Thus we have $\frac{V}{n} = d$

If $n = 1$, then volume V_m becomes

$$\frac{V}{n} = d \text{ or } V_m = d$$

where V_m is the molar volume (i.e. the volume of 1 mole of a gas)

Under standard temperature and pressure (STP) the volume occupied by 1 mole of a gas has been found to 22.4 L.

The Avogadro's law can be stated, "*Equal volumes of all gases at the same temperature and pressure contain the same number of molecules*".

Since one mole of anything (solid, liquid or gas) always contains 6.022×10^{23} particles (Avogadro's number). So in 22.4 litre of any gas at NTP the number of molecules is 6.022×10^{23}.

Example 1. *How many formaldehyde (HCHO) molecules are there in a sample of the gas which occupies 5.6 L at S.T.P?*

Solution.

Number of moles of formaldehyde

$$= \frac{5.6 \times 1}{22.4} = 0.25$$

Number of formaldehyde molecules

$$= 0.25 \times 6.22 \times 10^{23} = \mathbf{1.506 \times 10^{23}}.$$

The Ideal Gas Equation (or Gas law)

The *ideal gas law* combines all the information for the Boyles law, Charles law and Avogadros law. The equation for this law is

$$PV = nRT$$

where R is molar gas constant. The value of R is

$$R = 0.08206 \text{ L-atm/(K.mol)}$$

This ideal gas equation can be used to predict the behaviour of a gas under changes of temperature and pressure.

All gases behave ideally provided that the pressures are relatively low and the temperature are relatively high.

A gas which behaves exactly as predicted by the gas laws or which obeys the ideal gas equation under all conditions of temperature and pressure is called an **Ideal gas.**

We can express the values of R in any set of units representing work and energy.

In C.G.S units work is expressed in ergs. The unit of R is ergs degree^{-1} mole $^{-1}$. Since 1 calorie = 4.184 x 10^7 ergs, R can also be expressed in calories degree^{-1}. If P is expressed in atmospheres and V in litres

$$R = \frac{\text{Atmosphere} \times \text{litre}}{\text{degree} \times \text{mole}}$$

$$= \text{litre-atmosphere degree}^{-1} \text{ mole}^{-1}$$

In S.I. systems of units, cubic decimeters are used instead of litres and Joules (J) are used instead of calories

$$1\text{ J} = \text{Newton (N)} \times \text{metre (m)}$$

$$R = \frac{\text{Newtons} \times \text{metres}}{\text{Moles} \times \text{Degrees}}$$

$$= \text{Nm}^{-1}\text{K}^{-1}\text{mol}^{-1}$$

[Note that, since Newton × metres (force × length) represents energy]

Numerical Values of *R*

One mole of any ideal gas at standard conditions (S.T.P) i.e. at 0°C and 1 atmosphere pressure, occupies a volume of 22.413 litres. If we express then the volume in litres and the pressure in atmospheres, then R can be expressed as

$$R = \frac{PV}{nT} = \frac{1 \times 22.413}{1 \times 273.15}$$

$$= \mathbf{0.08205} \text{ litre-atmospheres degree}^{-1} \text{ mole}^{-1}.$$

The **real gases** behave ideally if the pressure are low and temperatures relatively high. Real gases are often mixtures of pure gaseous substances and the gas laws stated above hold good for such mixtures under moderate conditions of temperature and pressure.

The gas equation $PV = nRT$ can also be written as

$$PV = \frac{W}{M} RT \qquad \left(\because n = \frac{W}{M}\right)$$

or

$$P = \frac{W}{V} \times \frac{RT}{M}$$

$$= d \cdot \frac{RT}{M} \qquad \left(\because d = \frac{W}{M}\right)$$

where W = weight of a given quantity of gas
and M = molar mass of the gas
d = density

The Significance or *R*

If we solve the equation $PV = nRT$ for R,

we get

$$\text{R} = \frac{PV}{nT} = \frac{\text{Pressure} \times \text{Volume}}{\text{Moles} \times \text{Temperature(K)}}$$

Since pressure is force per unit area and volume is area times length, it immediately follows that the units of PV/nT and hence of R are

$$R = \frac{\text{force} \times (\text{length})^3}{(\text{length})^2 \times \text{moles} \times \text{temperature(K)}}$$

$$= \frac{\text{force} \times \text{length}}{\text{moles} \times \text{temperature(K)}}$$

$$= \frac{\text{energy}}{\text{moles} \times \text{temperature(K)}} \left[\because \text{force} \times \text{length} = \text{energy}\right]$$

The values of R will therefore, be expressed in units of energy per Kelvin per mole.

If the pressure is expressed in atmospheres while the volume in cubic centimeters, R because,

$$R = \frac{1 \times 22.413}{1 \times 273.15} = \mathbf{82.05} \text{ C.C.-atm. degree}^{-1} \text{ mole}^{-1}.$$

If the pressure be expressed in dynes per square centimeter and volume in cubic centimeters, then

$$R = \frac{PV}{nT} = \frac{76 \times 13.6 \times 981 \times 22.413}{1 \times 273.15}$$

$$= 8.314 \times 10^7 \text{ ergs degree}^{-1} \text{ mole}^{-1}$$

Since 1 Joule = 10^7 ergs,
and 1 calorie = 4.184 Joules so

$$R = 8.314 \text{ Joules degree}^{-1} \text{ mole}^{-1}$$

$$= \frac{8.314}{4.184} = \mathbf{1.987\ Cal} \text{ degree}^{-1} \text{ mol}^{-1}.$$

The values of gas constant of gas occupies 0.0224 m^3 at 0°C and 101300 Pa pressure

$$R = \frac{PV}{T} = \frac{101300 \times 0.0224}{0 + 273}$$

$$= \mathbf{8.314\ Nm}\ K^{-1} \text{ mole}^{-1} \ (= JK^{-1} \text{ mol}^{-1})$$

Hence the various values of R are

0.082057 litre atm k^{-1} mol^{-1}

82.057 cm^3 atm k^{-1} mol^{-1}

8.31441 JK^{-1} mol^{-1}

8.31441 x 10^7 ergs k^{-1} mol^{-1}

1.98719 cal K^{-1} mol^{-1}.

Example 1. *Calculate the value of molar gas constant R in (i) ergs deg^{-1} mol^{-1} (ii) litre-atm deg^{-1} mol^{-1}.*

Solution. *(i)* In C.G.S. system P (1 atm) = cm Hg

$$= 76 \times 13.56 \times 980 \text{ dyn cm}^{-2}$$

$$V\ (22.4\ l) = 22400 \text{ cm}^3$$

$$n = 1.0 \text{ mol};\ T = 273 \text{ K}$$

$$R = PV/nT = \frac{76 \times 13.56 \times 980 \times 22400}{273}$$

$$= \mathbf{8.314 \times 10^7 \text{ erg } K^{-1} \text{ mol}^{-1}}$$

When the pressure is expressed in atmospheres and the volume in dm^3 (or *litres*). Knowing that 1 mole of a gas occupies a volume=22.4 dm^3 (or litres) at STP (273 K and 1 atmosphere pressure),

$$R = \frac{PV}{nT} = \frac{1 \times 22.4}{1 \times 273} = \mathbf{0.08205 \text{ litre - atm } K^{-1} \text{ mol}^{-1}}.$$

Example 2. *What volume would 1.0 g of Ar (mol. wt. = 40) occupy at 300 K and a pressure of 50 Nm^{-2} (R=8.314 JK^{-1}* mol^{-1}).

Solution.

Here $P = 50 \text{ Nm}^{-3};\ T = 300 \text{ K}$

$$R = 8.314\ JK^{-1} \text{ mol}^{-1}$$

$$\text{Amount of argon } (n) = \frac{\text{mass of substance(m)}}{\text{molar mass } (M)} \left(\because n - \frac{m}{M}\right)$$

$$= \frac{1.0}{40 \cdot 00} = 0.025 \text{ mole}$$

Substituting these values in the ideal gas law, we get

$$PV = nRT$$

or

$$V = \frac{nRT}{P} = \frac{0.025 \times 8.314 \times 300}{50}$$

$$= 1.247 \text{ m}^3.$$

Graham's Law of Diffusion or Effusion

To begin with let us make the terms *Diffusion* and *Effusion* clear.

If a gas is allowed to escape from its container through a small hole into vacuum, the rate at which molecules escape will depend upon the rate at which they reach the area represented by the hole. This process is called *Effusion*.

An analogous process is that of **diffusion** which *is the passage of gas through a porous partition such as a porous-pot.*

Diffusion is similar to *effusion* with a large number of holes through which the molecules escape instead of one single hole.

Graham's law of diffusion or effusion was enunciated by Graham and is based on his work on diffusion and effusion. It states, "*The rates of diffusion (or effusion) of gases under similar conditions of temperature and pressure are inversely proportional to the square roots of their densities.*"

Mathematically,

$$\frac{r_2}{r_1} = \sqrt{\frac{d_1}{d_2}} = \sqrt{\frac{M_1}{M_2}}$$

Effusion rates are most readily compared by taking two gases and allowing them in turn to pass through the same effusion hole, under identical conditions of temperature and pressure. Comparing the time for the same volume of each gas to pass through, it is found that

$$\frac{t_1}{t_2} = \sqrt{\frac{d_1}{d_2}} = \sqrt{\frac{M_1}{M_2}}$$

Graham's Law of diffusion is used in the separation of isotopes. Aston first used this method in 1913 for separation of isotopes of neon. In the recent times it is used for large scale separation of the isotopes of uranium. U-235 is separated by the repeated diffusion of uranium hexafluoride, which is gaseous at 60°C.

Example 1. *Time of diffusion of same volume of oxygen and chlorine under similar conditions are 25s and 37.5s. If the relative vapour density of oxygen is 16, calculate that of chlorine.*

Solution.

From the equation, $\frac{{}^{t}O_2}{{}^{t}Cl_2} = \sqrt{\frac{{}^{\rho}O_2}{{}^{\rho}Cl_2}}$

Substituting the given value of

${}^{t}O_2 = 25s$; ${}^{t}Cl_2 = 37.55$ and ${}^{\rho}O_2 = 16$, we get

$$\frac{25}{37.5} = \sqrt{\left(\frac{16}{{}^{\rho}Cl_2}\right)}$$

squaring both sides, we get

$$\frac{25 \times 25}{37.5 \times 37.5} = \frac{16}{{}^{\rho}Cl_2}$$

or

$$ {}^{\rho}Cl_2 = \frac{16 \times 37.5 \times 37.5}{25 \times 25} = 36$$

Hence vapour density of Cl_2 (${}^{\rho}Cl_2$) = **36.**

Dalton's Law of Partial Pressures

Gases exert pressure as any surface that they touch because the gas particles collide with the surface and push against it. For example, air exerts pressure on all sides of an exposed object because gas particles in air collide with it. In a mixture of gases, each gas exerts its own pressure, but how do these pressures relate to the total pressure exerted by the gaseous mixture?

According to John Dalton, the components of mixture of gases act independently in so far as their pressures are concerned, and the total pressure is the sum of the pressures due to the components. The pressure that each component of a gas mixture exerts is known as **Partial Pressure** of that component. According to **Dalton's Law of Partial Pressure,** the total pressure *exerted by a gaseous mixture is the sum of the partial pressures of the components of the mixture.* The total pressure (p) is given by

$$P = p_A + p_B + p_C + \ldots\ldots$$

where p_A, p_B, p_C are the partial pressures of components A, B, C etc.

This law exists because the **partial pressures** exerted by any one of the gases in the mixture is the same as it would be if the gas were alone.

If n_1 moles of a gas (A) are mixed with n_2 moles of another gas (B) in a flask of volume V and at temperature T. Partial pressures of the two gas (A and B) can be written as

$$p_A = \frac{n_1RT}{V} \text{ and } p_B = \frac{n_2RT}{V}$$

Total pressure of the mixture is

$$P = p_1 + p_2 = \frac{n_1RT}{V} + \frac{n_2RT}{V}$$

$$= \left(\frac{n_1 + n_2}{V}\right)RT.$$

Dalton's law of partial pressures is especially useful in calculations involving gases collected over water. Here the total pressure is equal to the vapour pressure of water (**aqueous tension**) plus the pressure due to the dry gas.

$$^{P}\text{Total} = {}^{P}\text{H}_2\text{O} + {}^{P}\text{gas}$$

Example 1. *A mixture of Hydrogen and Nitrogen enclosed in a 5 litre flask at 298 K exerts a pressure of 2.5 atmospheres. If the amount of hydrogen present is 0.30 moles, calculate the partial pressure of two gases.*

Solution.

Let the partial pressures of hydrogen and nitrogen be p_1 and p_2 respectively, then

Total pressure = $p_1 + p_2$ = 2.5 atoms pheres

Amount of H_2 present = 0.30 mole

$$p_1 = \frac{n_1RT}{V} = \frac{0.3\,\text{mol} \times 0.0821\,l\,\text{atm K}^{-1} \times 298\text{K}}{5\,l}$$

$= 1.468$ atm.

$p_2 = P - p_1 = 2.5 - 1.468 = 1.032$ atm

$\therefore$ Partial pressure of hydrogen=1.468 atm.

and Partial pressure of nitrogen=1.032 atm.

1.6 The Kelvin Scale or Absolute Scale of Temperature

Temperature is one of the basic quantities. In the measurement of temperature we use any property of a substance which depends upon and changes continuously and uniformly with temperature. Clearly the expansion of mercury in glass tube cannot be guaranteed to be a phenomenon on which to build a basic, simplifying description of the physical world. It is furthermore, unusual to make an arbitrary assignment of a zero value to the freezing point of water.

A better base for the development of a quantitative description of temperature is provided by the common expansion behaviour of gases. The volume temperature relationship for a gas is shown in Fig. 1.7. It can be seen that the straight lines that occur when w plot the volume of two different gases at constant pressure against their celsius temperatures. When these straight lines are extended from the last experimental point toward lower temperatures, something surprising is noticed. Both lines intersect the horizontal axis (where the graph indicates that the volume of the gas is zero) at the same temperature (i.e. –273°C). Since a negative volume is not physically possible, the experimental results suggest that this temperature is the lowest possible. In order to accommodate such a conclusion, we can create a new temperature scale in which the lowest possible temperature becomes the zero point. We do so by adding 273 to any celsius temperature. The resulting temperature scale is called the *Kelvin scale* in honour of

William Thompson, Lord Kelvin (1824-1907), the Scottish physicist and nobleman who first proposed it.

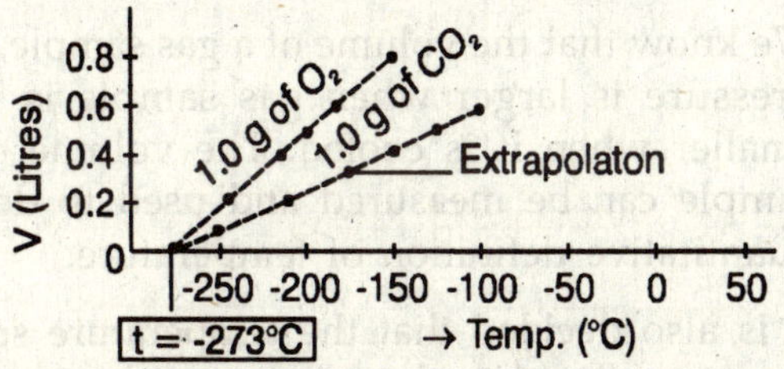

Fig. 1.7A. Volume-Temp. relationship for gases at constant pressure.

The relationship between the *Kelvin scale* and *Celsius scale* is

$$\boxed{K = {}^\circ C + 273}$$

When we replot the volumes of gases as in the figure above (Fig. 1.7 A) against Kelvin temper we get as graph as shown in figure below (Fig 1.7B)

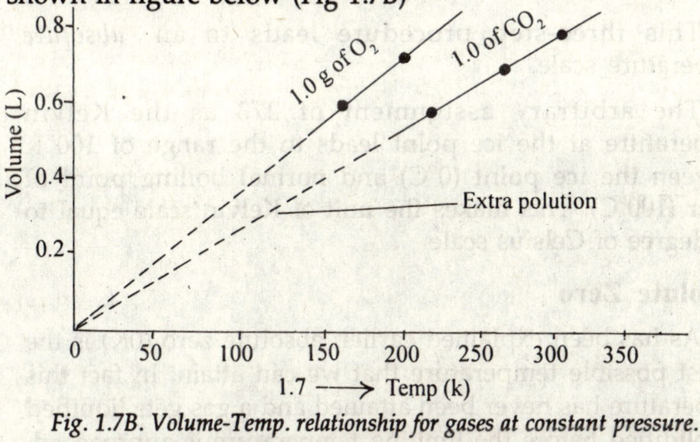

Fig. 1.7B. Volume-Temp. relationship for gases at constant pressure.

In this figure (Fig. 1.7 B) we find that the straight lines extrapolate to 0K, where the graph indicates that the volume is also equal to zero. From the graph it appears that the matter disappears at *absolute zero* (0K). However when real gases are cooled they liquify before their temperature reaches 0K (absolute zero).

This concept of "absolute zero" temperature can be developed by the following three steps. It is based on the common expansion behaviour of gases. The various steps are:—

(i) We know that the volume of a gas sample, at fixed pressure is larger when gas sample is hot and smaller when it is cooled. The volume of a gas sample can be measured and used to develop a quantitative definition of temperature.

(ii) It is also decided that the temperature scale will be proportional to the volume of the gas sample, at some fixed pressure. Thus the temperature scale will have a zero value when the volume of the gas sample has a zero value.

(iii) The temperatures on the *Kelvin scale* are obtained by assigning a value of 273 at the temperature of ice point (0°C).

This three-step procedure leads to an *absolute* temperature scale.

The arbitrary assignment of 273 as the Kelvin temperature at the ice point leads to the range of 100 K between the ice point (0°C) and normal boiling point of water (100°C). This makes the unit of Kelvin scale equal to the degree of Celsius scale.

Absolute Zero

As has been explained earlier, absolute zero (0K) is the lowest possible temperature that we can attain. In fact this temperature has never been attained and a gas gets liquified or solidified before the limiting temperature is approached. At this temperature, the volume of the gas (although theoretically expected to be zero) will not become zero. It would reduce to the volume occupied by the particles themselves and no space would be available in which the molecules could move. The volume occupied by the system under such conditions would be negligibly small as compared with the volume of the substance in the normal gaseous state.

2

Kinetic Molecular Theory of Gases

2.1 INTRODUCTION

Avogadro's law conceives molecules as discrete particles in a gas. *Brownian motion*, first observed by the botanist Robert Brown, in 1827, also points to the same conclusion. In gases the molecules move around rapidly. (The term *Kinetic* in *kinetic molecular theory* refers to this motions).

Because a gas consists of molecules within lots of empty space, a gas is compressible. Since the gaseous particles in *particle model* possess energy of motion, called *kinetic energy*, the mathematical expression describing this model is called the **Kinetic theory of gases.** Thus, the use of the molecular model to understand the quantitative properties of the gases is known as the *Kinetic molecular theory of Gases*. The work of Ludwig Baltzmann, James Clerk Maxwell, and R.J. Claussius during the late 1800s was mainly responsible for the development of this theory.

The *Kinetic molecular model* of an ideal gas is set up by the following statements:—

(i) A gas is made up of a large number of particles, or molecules, that are small in comparison with the size of the gas sample.

(ii) The gaseous molecules are the continuous, randomly directed motion.

(iii) The molecules are independent of each other and interact with each other or with the walls of the container only during brief collisions. These collisions, on the average, are perfectly elastic i.e. none of the translational energy is lost as a result of the collisions.

(iv) The equation $F = m \times a$, can be used to describe the average interaction of the molecules with the walls of the container.

2.2 Kinetic Theory of Gases

The *Kinetic molecular theory of gases,* developed during eighteenth and nineteenth century by various scientists provides an explanation for the behaviour of gases.

This theory was developed for *ideal gas*. Though ideal gas does not exist, all gases approach the ideal gas behaviour under moderate conditions of temperature and pressure.

The important of this theory are:

(i) A gas is made up particles (atoms or molecules) whose dimensions are very small compared with the very large average distance between the particles.

(ii) These particles are in constant random motions, moving in straight lines unless they collide with another particle or the walls of the container.

(iii) Attraction or repulsion between gas particles is negligible.

(iv) The average kinetic energy (the energy of motion) of these particles is indirectly proportional to the Kelvin temperature.

Although the (iii) postulate states that attraction between gas particles are negligible, attractions must exist or gases would never condense into the liquid state. Attractive forces are responsible for holding the liquid state together. Clearly the assumption that these forces are absent in gases is a simplification. Nevertheless, attractive forces between gas particles and negligible under moderate conditions of temperature and pressure.

According to (iv) postulate, the average kinetic energy of gas particles is directly proportional to the Kelvin temperature. The term directly proportional means that the average kinetic energy will double when Kelvin temperature is doubled. The motion of gas particles became more chaotic and violent as the temperature increases and more restrained as the temperature decreases.

These postulates explained the behaviour of gases as follows:

Gases fill a container completely and uniformly because the particle move constantly and rapidly and virtually have no attraction to one another. Gases are easily compressed and have low densities because of very large average distances between the particles. In fact most of the space in any gas sample, under moderate conditions is empty space (without any gas particles.) Finally gases exert a pressure on any surface that they touch because the gas particles collide with the surface and push against it.

2.3 Pressure, Volume and Temperature of a Gas in the Light of above Postulates

Pressure Exerted by a Gas

When a certain mass of gas is enclosed in a vessel, the molecules collide with the walls of the container and are being reflected from them. During the collision each molecule suffers a change in momentum.

According to Newton's second law of motion, the rate of change of momentum is equal to the force exerted. This pressure can be formed out using the postulates of kinetic theory of gases.

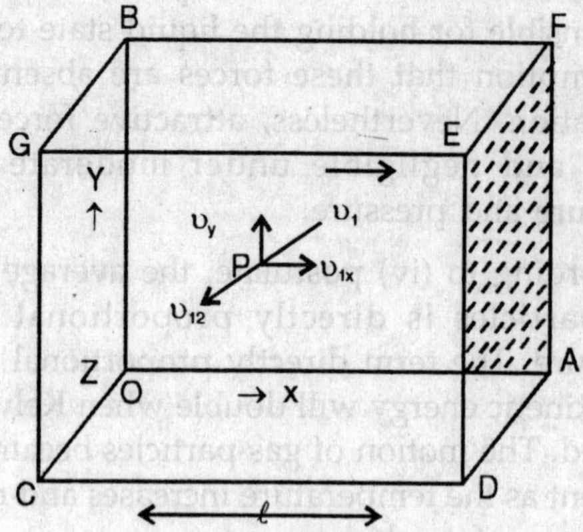

Fig. 2.1.

Consider a perfect gas enclosed in a cubical vessel of side '1' having perfectly elastic walls. The number of gas molecules enclosed in the vessel be *'n'* and the mass of each molecule be *'m'*. Let the sides of the cube represent mutually perpendicular X, Y and Z axes.

Consider a molecular P moving in random direction with a velocity υ_1. Let $\upsilon_{1\xi}$, υ_{1y} and v_{1z} be the components of υ_1 along three mutually perpendicular axes x, y and z, respectively. Then

$$\upsilon_1^2 = \upsilon_1^2 x + \upsilon_1^2 y + \upsilon_1^2 z$$

The component of velocity with which the molecule P will strike the face DFEA is υ_{1x} and the momentum of the molecule along x-direction before collision is mx υ_{1x}.

Since the collision is perfectly elastic so during collision there is no change in the speed of the molecule and only the direction of velocity changes. Hence

The momentum after collision, along x-axis is

$$= m\ (-\upsilon_{1x})$$
$$= -m\ \upsilon_{1x}$$

The change in momentum $= m\ \upsilon_{1x} - (-m\upsilon_{1x})$

$$= -\,2m\upsilon_{1x}$$

For simplicity we treat each component of velocity separately.

The molecule travels a distance of $2l$ before striking the same face again.

Thus the time between two successive collisions on the same face in $\frac{2l}{\upsilon_{1x}}$

Hence the number of collisions per second $\frac{\upsilon_{1x}}{2l}$

Thus the change in momentum per second on this face

$$= 2m\upsilon_{1x} \times \frac{\upsilon_{1x}}{2l}$$

$$= \frac{m \cdot \sigma_1^2 x}{l}$$

Since all the molecules of a gas are identical and the time spent in collision is negligible, we can safely assume that the original molecule continues to move with its initial speed. Considering all the n molecules striking the same face, we get total change in momentum.

$$\upsilon_1^2 x = \frac{m \cdot \upsilon_1^2 x}{l} + \frac{m \cdot \upsilon_2^2 x}{l} + \frac{m \cdot \upsilon_3^2 x}{l} + \ldots\ldots \frac{m \cdot \upsilon_n^2 x}{l}$$

$$= \frac{m}{l} \sum_i \upsilon_i^2 x$$

Where $\upsilon_{1x}, \upsilon_{2x}, \ldots \upsilon_{nx}$ and x-components of the velocities $\upsilon_1, \upsilon_2 \ldots \upsilon_n$ of the n molecules.

Since rate of change of momentum is equal to the force exerted by the gas, so the net force exerted by the gas on this face is equal to

$$= \frac{m}{l} \sum_i \upsilon_i^2 x$$

Hence Pressure P is given by

$$P_x = \frac{\text{Force}}{\text{Area}}$$

$$= \frac{m}{l \cdot l^2} \sum_i u_i^2 x$$

or

$$P_x = \frac{m}{V} \sum_i u_i^2 x$$

Similarly we can get

$$P_y = \frac{m}{V} \sum_i u_{iy}^2$$

and

$$P_z = \frac{m}{V} \sum_i u_{iz}^2$$

When the cube in small, the gravity does not play any role and the pressure on all faces may be taken as equal in magnitude

i.e. $P_x = P_y = P_z$

By adding we get

$$3P = P_x + P_y + P_z = \frac{m}{V} \sum_i \left(\upsilon_{ix}^2 + \upsilon_{iy}^2 + \upsilon_{iz}^2\right)$$

or

$$P = \frac{1}{3} \frac{m}{V} \sum_i \left(\upsilon_{ix}^2 + \upsilon_{iy}^2 + \upsilon_{iz}^2\right)$$

$$= \frac{1}{3} \frac{m}{V} \sum_i \upsilon_i^2$$

Now if υ^2 is the mean average speed of all the molecules, then

$$\upsilon^2 = \frac{V_1^2 + V_2^2 + \ldots V_n^2}{n} = \frac{\sum_i \upsilon_i^2}{n}$$

or

$$\sum \upsilon_1^2 = n\upsilon^2$$

Thus

$$P = \frac{1}{3} \frac{mn\overline{\upsilon}^2}{V}$$

But $m\ n$= M= Total mass of the gas

So $\frac{mn}{V} = l$ = density of the gas

Therefore pressure of the gas becomes

$$P = \frac{1}{3} l \overline{\upsilon}^z$$

or

$$E = \frac{2}{3} \cdot \frac{1}{2} l \overline{\upsilon}^2 = \frac{2}{3} E$$

where

$$E = \frac{1}{2} l \overline{\upsilon}^2 = \text{K.E. per unit volume.}$$

Thus *the pressure of a gas is equal to two-third of total translational K.E. of the molecules per unit volume.*

2.4 Some Deduction from Expression for Pressure

If 1 gram molecule of a gas are contained in a Volume V at Temperature T absolute, then

$$P = \frac{1}{3} \frac{mN}{V} \overline{\upsilon}^2$$

where

N = Avagadro's Number.

Volume: The volume of a gas is mostly empty space in which billions of tiny molecules are in rapid motion.

Temperature: The temperature of a gas is a measure of the average kinetic energy of its molecules. Greater the kinetic energy higher will be the temperature.

Kinetic Interpretation of Temperature

From the kinetic equation, we have for 1 mole of a gas

$$PV = \frac{1}{3} mN \overline{\upsilon}^2$$

According to ideal gas equation, for 1 mole

$$PV = RT$$

Comparing these two equations, we get

$$\frac{1}{3}mN\overline{\upsilon}^2 = RT$$

or
$$\frac{1}{2}m\overline{\upsilon}^2 = \frac{3}{2}\frac{R}{N}T$$
$$= \frac{3}{2}kT$$

Where $k = \frac{R}{N}$ = Bltzmann constant = 1.38×10^{-23} J/K

From the above equation we observe that *Average kinetic energy of translation of molecules is directly proportional to the absolute temperature of the gas.*

Thus average kinetic energy of the molecules can be considered as a measure of temperature.

Moreover at absolute zero, the kinetic energy of gas molecules must be zero i.e. the molecular mass of the gas ceases. Hence **absolute zero** is that temperature at which all molecular motion of a gas ceases.

From the above equation we also have

$$\overline{\upsilon}^2 \propto T$$

i.e. *the mean square velocity of the molecules is directly proportional to absolute temperature.*

2.5 Validity of the Postulates of the Kinetic Theory of Gases

(*i*) Extreme compressibility of a gas is proof of the large empty space in it [*Postulate* (*i*)]. It has been calculated that 99.96% of the total volume of oxygen at NPT is empty space.

(*ii*) Dust particles in air are found to be in constant irregular zig-zag motion termed *Brownian motion*. This implies the very fast straight line motion of the molecules [*Postulate* (*ii*)].

(*iii*) Rapid motion of so many molecules must involve frequent collisions between them. These collisions must be elastic [*Postulate* (*iii*)] as otherwise there will be constant loss of kinetic energy and ultimately the molecules must come to rest (something which does not happen).

(*iv*) In a compressed gas the molecules are very close and intermolecular attraction is the highest. The attraction must be negligible [*Postulate* (*iv*)] as even a highly compressed gas expands freely and fills the whole available space.

(*v*) Diffusion is an inherent property of gases. Even a heavier gas rises up to intermix with a lighter gas. This suggests that the effect of gravity is negligible in comparison to the effect of continued collisions between them [*Postulate* (*v*)].

(*vi*) On heating, the temperature of the gas rises. Heat energy supplied increases the speed of the molecules and their average kinetic energy increases resulting in rise of temperature [*Postulate* (*vi*)].

Kinetic Energy of an Ideal Gas: The molecules of a gas are in a state of constant rapid motion and possess kinetic energy due to their motion. The total kinetic energy of translation of the molecules present in one mole of any gas can be calculated with the help of the equation

$$PV = \frac{1}{2} mNc^2$$

where m is the mass of the molecule of the gas possessing root mean square velocity c and N is the number of molecules of the gas.

This equation is known as Kinetic equation for ideal gases. We shall derive it from Kinetic Theory of Gases in the higher classes.

But $$PV = RT \text{ (Gas equation for one mole of an ideal gas)}$$

For of mole of a gas $N = N_A$ (Avogadro's number)

$$\therefore \quad \frac{1}{2} mN_A c^2 = RT \quad \ldots(i)$$

or $$\frac{2}{3}\times\frac{1}{2}mN_Ac^2 = RT$$

But $\frac{1}{2}mc^2$= average kinetic energy of one individual molecule, $\overline{E}_k$. Hence total kinetic energy per mole of N_A molecules present in the sample

$$E_k = N_A\overline{E}k = \frac{1}{2}mN_AC^2 \qquad ...(ii)$$

From eqn. (*i*) and (*ii*)

$$\frac{2}{3}E_k = RT$$

where E_k, the total kinetic energy for 1 mole of a gas

$$k\varepsilon = \frac{1}{2}Mc^2 = \frac{3}{2}RT \qquad ...(iii)$$

Total kinetic energy for *n* moles of the gas $E_k = \frac{3}{2}nRT$

Since the equation does not involve any *P* and *V* terms, it is clear that *mean kinetic energy of an ideal gas is independent of the pressure and the nature of the gas and is proportional to the absolute temperature.* This is also the **Maxwell's generalisation** which states that at a given temperature the mean kinetic energy *E* of an ideal gas in independent of the nature of the gas. From eqn. (*iii*) it is seen that

$$c_2 \propto T \text{ or } c \propto \sqrt{T}$$

or *c* which is the measure of molecular motion is proportional to square root of the absolute temperature of the gas. That is why the molecular motion is termed the thermal motion also.

Example 1. *Calculate the internal energy of one mole of nitrogen at 423K assuming it to be an ideal gas.*

Solution. Total kinetic energy of translation of molecules present in one mole of a gas is given by the formula

$$K.E. = \frac{3}{2}RT$$

Substituting the values of

$R = 8.315\ KJ^{-1}\ mol^{-1}$ and $T = 423$ K

we get $K.E = \frac{3}{2} \times 8.314 \times 423 = 5.277 \times 10^3$ J

Calculation of Molecular Velocities. From the kinetic gas equation

$$PV = \frac{1}{3}\ mNc^3$$

For one mole of a gas $PV=RT$

and also $PV = \frac{1}{3}mNc^2$ (where N is Avogadro number)

$$= \frac{1}{3}Mc^2 \quad (\because M, \text{ mass of 1 mole of gas} = mN)$$

Hence $\frac{1}{3}Mc^2 = PV=RT$

$$\text{Whence } C^2 = \frac{3PV}{M} = \frac{3RT}{M} = \frac{3P}{e} \qquad \left(\because \frac{M}{V} = \rho\right)$$

or $$C = \sqrt{\frac{3PV}{M}} = \sqrt{\frac{3RT}{M}} = \sqrt{\frac{3P}{e}}$$

Any one of the three formulate can be used to calculate the value of C, the root mean square velocity (RMS).

2.6 Kinetic Theory of Gases and Gas Laws

In science a theory is accepted only if it explains the experimentally observed facts, various gas laws are nothing but generalised statements of the behaviour of gases. Let us see if the kinetic theory of gases can successfully explain various gas laws or it fails to explain them.

According to kinetic theory of gases, at all temperatures, the molecules of a gas are in a state of rapid and random motion, colliding with one another and also with the walls of the container. Each time a molecule hits the wall, it gives it a push. As the walls are rigid and can not move so it experiences a force. This force per unit area is the pressure of the gas. Temperature of the gas is a measure of kinetic energy of its molecules. We also know that average kinetic energy of the gas molecules is directly proportional to temperature on Kelvin scale.

Average kinetic energy, $\overline{E}_k \propto T$

EXPLANATION OF VARIOUS GAS LAWS

Boyle's Law

The kinetic molecular theory of gases provides an explanation for Boyle's law. According to (ii) postulate of the theory, gas molecules are in constant, random motion, moving in straight line, until they collide with another particle or the walls of the container. Since average kinetic energy is directly proportional to Kelvin temperature, so the average energy of collision when one gas molecule hits the wall of the container will not change when volume decreases at constant temperature. However the pressure will increase when the volume decrease because there will be more collisions with the walls of the container per unit area. This is Boyles's Law.

Also from kinetic equation of gas, we have

$$PV = \frac{1}{3}mN\overline{\upsilon}^2$$

Since

$$\frac{1}{2}m\overline{\upsilon}^2 \alpha T$$

So at constant T, we have

$$PV = \frac{2}{3}N \cdot \frac{1}{2}m\upsilon^z = \frac{2}{3}N\alpha T$$

This is Boyles's Law.

Charles's Law

The kinetic molecular theory explains Charles Law.

Consider a gas contained in an elastic container. On cooling the pressure inside the container will not change because the pressure of the gas inside an elastic container is always identical to atmospheric pressure. However due to decrease in temp, the average kinetic energy of the gas particles will decrease. Less powerful collisions with the walls of container will occur since the gas particles exert less pressure on the walls of the container and to keep the pressure equal to atmospheric pressure the area of the walls of container must decrease if the collision force decreases, hence the container must contract i.e. its volume must decrease. This is Charles Law.

From the kinetic gas equation

$$PV = \frac{1}{3}mN\bar{\upsilon}^2$$

We know, $K.E. = \frac{1}{2}m\bar{\upsilon}^2 \alpha T$

$$\therefore \quad PV = \frac{2}{3}N\alpha T$$

$V \alpha T$ for a constant value of P. This is Charles Law.

Gas Equation

$$PV = \frac{1}{3}m N\bar{\upsilon}^2$$

Also $\frac{1}{2}m\bar{\upsilon}^2 \alpha T$

or

$$PV = \frac{2}{3}N \cdot \frac{1}{2}m\bar{\upsilon}^2 = \frac{2}{3}N\alpha T$$

or

$$PV = RT$$

Where $R = \frac{2N\alpha}{3} = N \cdot k = R.$

AVOGADRO'S HYPOTHESIS

Avogadro's Law is explained by (i) postulate of kinetic molecular theory, which says that dimensions of gas particle are very small compared with the very large average distances between them.

Consider two gases A and B at a pressure P and each having a volume V. Let m_1 and m_2 be the masses of each molecule of A and B respectively and υ_1 and υ_2 be the number of molecules of A and B respectively. If $\overline{\upsilon}_1^2$ and $\overline{\upsilon}_2^2$ be the mean square speed of A, B respectively. Then

$$\text{Pressure of A is } P = \frac{1}{3}\frac{m_1 n_1 \overline{\upsilon}_1^2}{V}$$

$$\text{and Pressure of } B \text{ is } P = \frac{1}{3}\frac{m_2 n_2 \overline{\upsilon}_2^2}{V}$$

From these two equation, we have

$$m_1\, n_1\, \overline{\upsilon}_1^2 = m_2\, n_2\, \overline{\upsilon}_2^2 \qquad ...(i)$$

If the two gases are at the same temperature (T), they must have the same kinetic energy.

i.e.

$$\frac{1}{2}m_1\overline{\upsilon}_1^2 = \frac{1}{2}m_2\overline{\upsilon}_2^2$$

or

$$m_1\overline{\upsilon}_1^2 = m_2\overline{\upsilon}_2^2 \qquad ...(ii)$$

Thus $n_1 = n_2$ [Comprising (*i*) and (*ii*)]

This is Avagadro's hypothesis.

GRAHAM'S LAW OF DIFFUSION

In accordance with kinetic theory, we have

$$\text{P} = \frac{1}{3}e\cdot\overline{\upsilon}^2$$

or

$$\text{v}_{ims} = \sqrt{\overline{\upsilon}^2} = \sqrt{\frac{3P}{e}}$$

i.e. root mean square velocity of the molecules of a gas is universally proportional to the square root of its density.

Consider two gases with root mean square velocities υ_1 and υ_2 respectively

Then

$$\upsilon_1 = \sqrt{\frac{3P}{e_1}}$$

$$\upsilon_2 = \sqrt{\frac{3P}{e_2}}$$

or

$$\frac{\upsilon_1}{\upsilon_2} = \sqrt{\frac{e_2}{e_1}}$$

As rate of diffusion of gas is proportional to the root mean square speed of molecules, thus from above equation it follows that the rate of diffusion of a gas is inversely proportional to the square root of its density.

This is Graham's Law of diffusion.

Dalton's Law of Partial Pressures

Let p_1 and p_2 be the partial pressures of the individual gases in a mixture of volume V_1. Let n_1 molecules of gas with partial pressure p_1 be present in the mixture. Let the average square velocity of these molecules be $\overline{\upsilon}_1^2$ with each having a mass m_1. Let the mixture contain n_2 molecules of the gas with partial pressure p_2. Let the mass of its molecules be m_2 and their mean square velocity be $\overline{\upsilon}_2^2$.

As each gas occupies the whole volume V at its own pressure, therefore

$$p_1 V = \frac{1}{3} m_1 \, n_1 \, \overline{\upsilon}_1^2$$

$$p_2 V = \frac{1}{3} m_2 \, n_2 \, \overline{\upsilon}_2^2$$

So that $(p_1+p_2+....)\ V = \frac{1}{3}\left(m_1 n_1 \overline{\upsilon}_1^2 + m_2 \overline{\upsilon}_2^2 + ...\right)$

Since the gaseous mixture is at the same temperature so the kinetic energy of translation per molecule of each gas is the same i.e.

$$\frac{1}{2} m_1 \overline{\upsilon}_1^2 = \frac{1}{2} m_2 \overline{\upsilon}_2^2 + ...$$

$$= \frac{1}{2} m \overline{\upsilon}^2$$

From the above two equations, we get

$$(p_1+p_2+....)\ V = \frac{1}{3}\ (n_1 + n_2....)\ m\overline{\upsilon}^2$$

But $(n_1 + n_2 + ...)$ is the total number of molecules in the mixture. Hence net pressure P is given by

$$PV = \frac{1}{3}\ (n_1 + n_2 + ...)\ m\overline{\upsilon}^2$$

or $\quad P = p_1 + p_2 + ...$ (By comparing above equation)

i.e. The net pressure due to a mixture of non-reacting gases contained in the same vessel is equal to the sum of the pressures which each gas would exert if it were alone contained in the vessel. This is Dalton's law of partial pressures.

The *Reasons for upper atmosphere being Cool* in spite of the fact that its kinetic temperature is of the order of 1000K.

In the upper atmosphere the heat content of an individual molecule increases by about 3.5 times as compared to that at the earth's surface. Kinetic temperature on earth's surface is about 300K whereas in the upper atmosphere it is of the order of 1000K. But the pressure and hence the density is very low in upper atmosphere i.e. in the upper atmosphere the number of molecules per C.C. is very small. Hence the total heat content in the upper atmosphere is very small as compared to the heat content on the earth in the same volume of the gas. This results in very low temperature of the upper atmosphere.

For pressure P_1, temperature T_1 and volume V_1, gas eqn. is

$$P_1V_1 = RT_1.$$

$$\therefore \quad R = \frac{PV}{T} = \frac{P_1V_1}{T_1} \text{ or } V_1 = \frac{PVT_1}{P_1T}$$

$$\therefore \quad V_1 = \frac{76 \times 13.6 \times g \times 22400 \times 300}{10^{-11} \times 13.6 \times g \times 273} = \frac{76 \times 22400 \times 300}{10^{-11} \times 273}$$

$$= 1.87 \times 10^{16} \text{ c.c.}$$

$$\therefore \text{ number of molecules per c.c.} = \frac{6 \times 10^{23}}{1.87 \times 10^{16}}$$

$$= 3.200 \times 10^7.$$

2.7 Maxwells Law of Distribution of Velocities

Kinetic theory of gases assumes that the molecules of a gas are moving continuously in all directions with different velocities and are colliding with one another. The velocity changes after each collision. Since collisions are perfectly elastic no loss of K.E. takes place. Thus after collisions some molecules will be speeded up and some will be slowed down. It means that even at constant temperature the molecular chaos exists.

In 1860 Maxwell pointed out that inspite of this molecular chaos there is one regularity. He suggested that *at any particular temperature the fraction of molecules possessing particular velocity remains almost constant.* This is known as 'steady state'. Considering the probability laws be calculated the fraction of molecular having particular velocity at an particular temperature.

On plotting the fraction of molecules against the corresponding velocities the curve obtained was always the same as shown in Fig 2.2. This curve is known as Maxwell's distribution of velocities. From the curve it can be easily observed:

(*i*) The fractions of molecules with very low or very high velocities are very small.

(*ii*) There is a velocity (α) corresponding to which the fraction of molecules is possessed by maximum number of molecules. It is known as *most probable velocity*.

The most probable velocity may be defined as the velocity possessed by a maximum fraction of molecules of any gas at a particular temperature.

It should be noted clearly that through the fraction of molecules corresponding to particular velocities remains constant at any particular temperature, the individual molecules are not the same at different instances of time. This is due to rapid collisions taking place in molecules continuously.

Effect of temperature on Maxwell's distribution of velocities

On increasing the temperature curve shifts as shown in Fig. 2.2. The important changes that can be noticed are:

(*i*) The peak has shifted forward.

(*ii*) The peak has shifted downward and flattened.

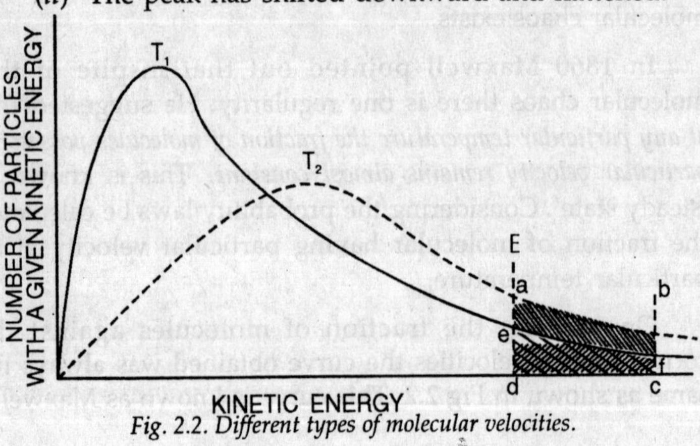

Fig. 2.2. Different types of molecular velocities.

From this the following conclusions may be drawn:

(*i*) On increasing temperature, the value of most probable velocity increases.

(*ii*) The fraction of molecules possessing the most probable velocity decreases with increase of temperature.

(*i*) **Root Mean Square Velocity (RMSV).** It may be defined *as the square root of the mean of the velocities of the molecules.* If v_1, v_2......etc., denote the velocity possessed by different molecules, then

$$C_{rms} \text{ or } \sqrt{\overline{C}^2} = \sqrt{\frac{v_1^2 + v_2^2 + v_N^2}{N}}$$

The expression to calculate this velocity can be derived from kinetic gas equation, according to which

$$PV = \frac{1}{3} mN_0\overline{C}_2$$

For 1 mol of the gas, $PV = RT$

$$\text{Thus} \quad \sqrt{\overline{C}^2} = c_{rms} = \sqrt{\frac{3PV}{mN_0}} = \sqrt{\frac{3RT}{mN_0}} = \sqrt{\frac{3RT}{M}}$$

$$= \sqrt{\frac{3kT}{m}} = \sqrt{\frac{3P}{d}}$$

where k denotes Boltzmann constant, $k = R/N_0$.

(*ii*) **Average velocity ($\overline{C}$).** It may defined *as the average of the various velocities possessed by the molecules, i.e.,*

$$\overline{C} = \frac{v_1 + v_2 + \ldots + v_N}{N}$$

It may be calculated by using the expression $\overline{C} = \sqrt{\frac{8kT}{\pi M}}$

$$\overline{C} = \sqrt{\frac{8RT}{\pi M}} = \sqrt{\frac{8kT}{\pi M}}$$

(iii) **Most Probable Velocity (MPV).** It may be defined as *the velocity corresponding to the maximum fraction of the molecules.* It may be given by the expression,

$$C_{mp} \text{ or MPV } = \sqrt{\frac{2RT}{M}} = \sqrt{\frac{2PV}{M}} = \sqrt{\frac{2P}{d}}$$

The relation between these velocities may be easily worked out

$$\frac{c_{mp}}{c_{rms}} \text{or} \frac{\text{MPV}}{\sqrt{C}} = \frac{\sqrt{2RT/M}}{\sqrt{8RT/\pi M}} = \sqrt{\frac{22}{28}} = \frac{1}{1.128}$$

$$\frac{c_{mp}}{c_{rms}} \text{or} \frac{\text{MPV}}{RMSV} = \frac{\sqrt{2RT/M}}{\sqrt{3RT/M}} = \sqrt{\frac{2}{3}} = \frac{1}{1.224}$$

Thus MPV: $\overline{C}$ RMSV:: 1: 1.128: 1.224.

i.e., RMSV > $\overline{C}$ > MPV.

This is also depicted in Fig. 2.3.

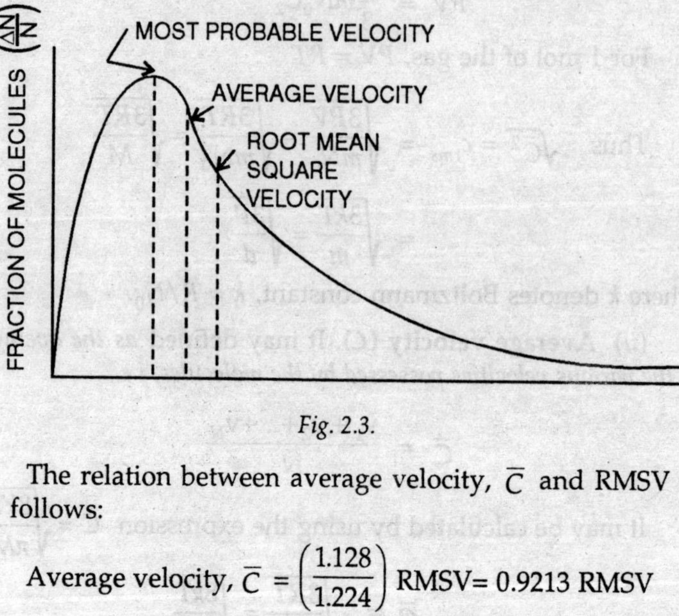

Fig. 2.3.

The relation between average velocity, $\overline{C}$ and RMSV is as follows:

$$\text{Average velocity, } \overline{C} = \left(\frac{1.128}{1.224}\right) \text{RMSV} = 0.9213 \text{ RMSV}$$

Rigorous derivation for c_{mp} and $\overline{C}$ gives

$$C_{mp} = \sqrt{\frac{2RT}{M}} = 1.4\sqrt{\frac{RT}{M}}$$

and $$\overline{C} = \sqrt{\frac{8RT}{\pi M}} = 1.6\sqrt{\frac{RT}{M}}$$

and $$C_{rms} = \sqrt{\frac{8RT}{\pi M}} = 1.7\sqrt{\frac{RT}{M}}$$

Hence the relative values of these velocities are

$$C_{rms}: \overline{C}\ C_{mp} :: 1.7: 1.6: 1.4.$$

Q. 1. *Calculate the root mean square velocity of oxygen molecule at 25°C.*

Solution. $$C_{rms} = \sqrt{\frac{3RT}{M}}$$

Here $T = 273 + 25 = 298$ K

$M = 32$

$R = 8.314 \times 10^7$ ergs deg^{-1} mole^{-1}

$$\therefore \quad C_{ms} = \sqrt{\frac{3 \times 8 \cdot 314 \times 10^7 \times 298}{32}}$$

Q. 2. *Calculate root mean square velocity of CO at 200°C.*

Solution. Here R = $8{\cdot}314 \times 10^7$ erg deg^{-1} mole^{-1}

$M = 28$

$T = 273 + 200 = 473$

Now $$C_{rms} = \sqrt{\frac{3RT}{M}}$$

$$= \sqrt{\frac{3 \times 8 \cdot 314 \times 10^7 \times 473}{28}}$$

$= 6{\cdot}49 \times 10^4$ cm/sec. **Ans.**

Q. 3. *Calculate the root mean square velocity of oxygen at N.T.P.*

Solution. Here $P = 1 \text{ atm.} = 76 \times 13.6 \times 981 \text{ dynes/cm}^2$

$$V = 22400 \text{ c.c.}$$

$$M = 32$$

Now $$C_{rms} = \sqrt{\frac{3PV}{M}}$$

$$= \sqrt{\frac{3 \times 76 \times 13.6 \times 981 \times 22400}{32}}$$

$$= 4.616 \times 10^4 \text{ cm/sec.}$$

Q. 4. *Oxygen has a density of 1.429 g. per litre at N.T.P. Calculate the root mean square and average velocity of oxygen molecules.*

Solution. $$C_{rms} = \sqrt{\frac{3PV}{M}}$$

$$= \sqrt{\frac{3P}{d}}$$

Here $$p = 1 \text{ atm.}$$

$$= 76 \times 136 \times 981 \text{ dynes/cm}^2$$

$$d = 1.429 \text{ g/litre}$$

$$= \frac{1.429}{1000} \text{g/ml}$$

$$= 0.001429 \text{ g/ml}$$

$\therefore$ $$C_{rms} = \sqrt{\frac{3P}{d}}$$

$$= \sqrt{\frac{3 \times 76 \times 13.6 \times 981}{0.001429}}$$

$$= 4.6 \times 10^4 \text{ cm/sec.}$$

Now Average velocity,

$$\overline{C} = 0.9213 \times C_{rms}$$

$$= 0.99213 \times 4.6 \times 10^4$$

$$= 4.2458 \times 10^4 \text{ cm/sec.}$$

2.8 CALCULATIONS OF DIFFERENT VELOCITIES FROM MAXWELLS DISTRIBUTION LAW

Maxwell-Boltzmann distribution is used for calculating the root mean square speed, the average speed and the most probable speed.

(*a*) **Root Mean Square Speed, C_{rms}:** In order to obtain the root mean square speed from the distribution expression, one has to multiply the number of molecules that have a particular value of the quantity to be averaged by that value of quantity. Then one will sum or integrate over all possible values of this quantity and ultimately divides by the total number of molecules in the sample. Hence C_{rms} is given as follows:

$$C_{rms} = \left(\bar{c^2}\right)^{1/2} = \left[\frac{1}{n'} - \int_0^{\infty} c^2 dn_e\right]^{1/2}$$

$$\frac{dn_e}{n'} = 4\pi\left(\frac{m}{2\pi kT}\right)^{3/2} \exp\left(-mc^2 / 2kT\right)c^2 dc \qquad \text{...(1)}$$

where m = mass of a gas molecule

T = temperature

k = Boltzmann constant and

$\frac{dn_c}{n'}$ = Fraction of the total number of molecules having speeds between c and $c+dc$

On substituting dn_c by the expression (1), we obtain

$$C_{rms} = \left(\bar{c}^2\right)^{1/2} = \left[4\pi\left(\frac{m}{2\pi kT}\right)^{3/2} \int_0^{\infty} c^4 \exp\left(-\frac{mc^2}{2k\text{T}}\right)dc\right]^{1/2}$$

But $\quad a = \frac{m}{2k\text{T}},$

$$\therefore \quad c_{bms} = \left(\bar{c}^2\right)^{1/2} = \left[4\pi\left(\frac{m}{2\pi kT}\right)^{3/2} \int_0^{\infty} c^4 \exp\left(-ac^2\right)dc\right]^{1/2}$$

But $\int_0^\infty c^4 e^{-ac^2} dc = \frac{3\sqrt{\pi}}{8a^{5/2}}$

Therefore the above equation becomes as follows:

$$c_{rms}(\bar{c}^2)^{1/2} = \left[4\pi\left(\frac{m}{2\pi kT}\right)^{3/2}\frac{3}{8}\frac{\sqrt{\pi}}{\left(\frac{m}{2kT}\right)^{5/2}}\right]^{1/2}$$

$$= \left[4\pi\left(\frac{m}{2\pi kT}\right)^{3/2}\frac{3}{8}\sqrt{\pi}\left(\frac{2kT}{m}\right)^{5/2}\right]^{1/2}$$

$$= \sqrt{\frac{3kT}{m}}$$

$$= 1.7\sqrt{\frac{RT}{M}} \qquad \ldots(2)$$

(*b*) **Average Speed:** In a similar way, one can calculate the average speed as

$$\bar{c} = \frac{1}{n}\int_0^\infty c\,dn_c$$

On putting the value of dn_c from equation (1), we get

$$\bar{c} = 4\pi\left(\frac{m}{2\pi kT}\right)^{3/2}\int_0^\infty c^3 \exp\left(-\frac{mc^2}{2kT}\right)dc$$

But $$a = \frac{m}{2kT}$$

Therefore, $\bar{c} = 4\pi\left(\frac{m}{2\pi kT}\right)^{3/2}\int_0^\infty c^3 \exp(-ac^2)dc$

Also $\int_0^\infty c^3 \exp(-ac^2)dc = \frac{1}{2a^2}$

Therefore, $\bar{c} = 4\pi\left(\frac{m}{2\pi kT}\right)^{3/2} \frac{4k^2T^3}{2m^2}$

or $$\bar{c} = \sqrt{\frac{4kT}{\pi M}}$$

$$= 1.6\sqrt{\frac{RT}{M}} \quad \text{...(3)}$$

(*c*) **Most Probable Speed, c_{mp}**: The probable speed is, obtained by differentiating the function on the right hand side of equation (1) with respect to *c* and equating the result to zero, i.e. we are applying the condition for maximum or minimum.

Hence,

$$2c\exp\left(-\frac{mc^2}{2kT}\right) + c^2\exp\left(-\frac{mc^2}{2kT}\right)\left(-\frac{2mc}{2kT}\right) = 0$$

$$2c\exp\left(-\frac{mc^2}{2kT}\right) + \frac{mc^2}{kT}\exp\left(-\frac{mc^2}{2kT}\right) = 0$$

$$c\exp\left(-\frac{mc^2}{2kT}\right)\left[2 - \frac{mc^2}{kT}\right] = 0$$

There are two possible solutions of this equation for *c* to be 0 and ∞ corresponding to the minima because molecules are not having speed equal to zero or infinity. The other solution can be found by putting the quantity in bracket, *i.e.*, $2-\frac{mc^2}{kT}$ equal to zero. This condition determines c_{mp}, *i.e.*,

$$2 - \frac{mc^2{}_{mp}}{kT} = 0$$

$$C_{mp} = \sqrt{\frac{2kT}{m}} = \sqrt{\frac{2RT}{M}}$$

$$= 1.4\sqrt{\frac{RT}{M}} \qquad ...(4)$$

From equations (2), (3) and (4) we get the ratio as

$$c_{rms}: c: c_{mp} = 1.7 : 1:6 : 1.4$$

$$1.0 : 0.92 : 0.82$$

Q. 1. *The compressibility factor, Z = PV/nRT for CO_2 at 273 K and 100 atm (101.325 × 10^5 N m^{-2}) pressure is at 100 atm and 273 K (a) by the ideal gas equation and (b) by making use of the compressibility factor.* **(Bombay 1988)**

Solution: (*a*) For an ideal gas

$$PV = nRT$$

or $$V = \frac{nRT}{P}$$

$$= \frac{(0.1 \text{ mol})(8.314 \text{ N m K}^{-1} \text{ mol}^{-1})(273.0 \text{ K})}{(101.325 \times 10^5 \text{ N m}^{-2})}$$

$$= 22.4 \times 10^{-6} \text{ m}^3$$

(*b*) Since, $$Z = \frac{PV}{nRT}$$

$$V = \frac{ZnRT}{P}$$

$$= \frac{(0.2007)(0.1 \text{ mol})(8.314 \text{ N mK}^{-1} \text{ mol}^{-1})(273.0 \text{ K})}{(101.325 \times 10^5 \text{ Nm}^{-2})}$$

$$= 4.5 \times 10^{-6} \text{ m}^3$$

Q. 2. *Oxygen at 1 atmosphere pressure and 0°C has density 1.4290 g/l. Find R.M.S. velocity of oxygen molecule.*

Solution. $$\sqrt{\frac{3PV}{M}} = \sqrt{\frac{3P}{D}}$$

$$\text{R.M.S. velocity} = \sqrt{\frac{3 \times 76 \times 13.6 \times 981}{0.001429}}$$

$$= 461.30 \text{ cm sec}^{-1}.$$

2.9 COLLISION PROPERTIES

Collision diameter: The distance between the centres of molecules when they are close to undergo collision is known as **Collision diameter** (σ).

It can be calculated as follows:

$$\text{(i)} \quad \sigma = \left[\frac{m\bar{\upsilon}}{3\sqrt{2}\pi\eta}\right]^{\frac{1}{2}}$$

Where $\bar{\upsilon}$ = average velocity
m = man of the molecule
η = viscosity

$$\text{(ii)} \quad \sigma = \frac{9}{4}\eta \cdot b\sqrt{\frac{\pi}{\text{PTM}}}$$

where η = viscosity
b = Vander Waals Constant
R = gas constant
M = molecular man

2.10 MEAN FREE PATH

The path described by any molecule between two successive collisions may be called the *free path* and the average of such free paths as the *mean free path*. If α_1, α_2, . . . α_n are the successive free paths traversed in total time t, then

$$\alpha_1, + \alpha_2 + ... \alpha_n = e. t$$

where e- is the average speed of the molecules

If n is the total number of collisions in time t, then

$$\alpha = \frac{\alpha_1 + \alpha_2 + \alpha_3 + ...\alpha_n}{n}$$

$$= \frac{\bar{c}\cdot t}{n} \qquad (\alpha_1 + \alpha_2 + \ldots \alpha_n = \bar{c}\cdot t)$$

α can also be calculated using the relation

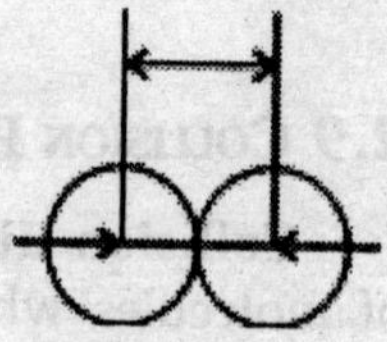

$$\alpha = \frac{1}{\sqrt{2}\pi\sigma^2 n}$$

where σ = collision diameter

N = Avogadro's number of molecules/cm^3

Also at constant P, mean free path α T

and at constant T, mean free path $\alpha \frac{1}{P}$

Also mean free path of a molecule is given by

$$\text{Mean free path} = \frac{3\eta}{e\,\upsilon_{rms}}$$

where η = coefficient of viscosity

e = density of gas

υ_{rms} = root mean square speed of molecules.

2.11 Collision Number

It is the number of collisions which a molecule registers with other molecules per second.

It can be calculated using the relation,

$$\text{Collision number} = \sqrt{2}\pi\sigma^2\bar{\upsilon}N$$

where $\bar{\upsilon}$ = average velocity

σ = collision diameter

N = number of molecules/cm^3.

2.12 Collision Frequency (Z)

It is the total number of collision taking place in a unit volume of the gas under particular conditions in one second.

It can be calculated using the relation

$$z = \frac{1}{2}\sigma^2\overline{\upsilon}N^2$$

where o = collision diameter

$\overline{\upsilon}$ = average velocity

N = number of molecules/cm^3.

Also at constant pressure,

collision frequency $\alpha\ T^{2/3}$

and at constant temperature

collision frequency $\alpha\ P^2$

2.13 Kinetic Energy of a Gas molecule

For 1 mole of a gas we have

$$PV = RT = \frac{1}{3}M\upsilon_{rms}^2$$

Kinetic energy for 1 gram molecule (mole) of a mono atomic gas is

$$\frac{1}{2}M\upsilon_{rms}^2 = \frac{3}{2}RT$$

$$K.E. = \frac{3}{2}RT$$

$$K.E. = \frac{3}{2}PV \qquad (\because PV = RT)$$

If the mass of each molecule is m and there are N numbers (Avogadro's number) of molecules then K.E. of each molecule,

$$\text{K.E. of each molecule} = \frac{3}{2}KT$$

Where K is Boltzmann's constant.

2.14 Boltzmann's Constant

The ratio of molar gas constant (R) to the Avogadro's

number N is called Boltzmann's constant.

i.e.
$$K = \frac{R}{N} = \frac{8.314\times10^{7}}{6.023\times10^{23}}$$
$$= 1.38\times10^{-23} \text{ Joules/degree}$$
$$= 8.167\times10^{-5}\ ev/k$$

∴ K.E. of a molecule is directly proportional to absolute temperature.

Equipartition of Energy. According to the principle of equipartition of energy, the average energy of each molecule (whether monoatomic, diatomic or polyatomic) in a gas associated with each degree of freedom is $\frac{1}{2}$ K.T.

2.15 Degree of Freedom

The total independent terms required to express the total energy of the molecule is called the *degree of freedom.*

(a) **Monoatomic gas.** The energy of each molecule in a monoatomic gas is expressed as

$$E = \frac{1}{2}m\upsilon_x^2 + \frac{1}{2}m\upsilon_y^2 + \frac{1}{2}m\upsilon_z^2$$

Here three independent variables $\upsilon_x, \upsilon_y, \upsilon_z$ are required to express its energy.

So the degree of freedom of a monoatomic gas molecule is three.

(b) **Diatomic gas.** The energy of each molecule of a diatomic gas is expressed as

$$E = \frac{1}{2}m\upsilon_x^2 + \frac{1}{2}m\upsilon_y^2 + \frac{1}{2}m\upsilon_z^2 + \frac{1}{2}I_x\omega_x^2 + \frac{1}{2}I_2\omega_z^2$$

Here the total independent terms required to explain its energy is five.

So the degree of freedom of a diatomic gas is five.

When the vibrational energy is taken into account the degree of freedom becomes seven.

(c) **Polyatomic gas.** The energy of each molecule of a polyatomic gas is expressed as

$$E = \frac{1}{2}mv_x^2 + \frac{1}{2}mv_y^2 + \frac{1}{2}mv_z^2 + \frac{1}{2}I_x\omega_x^2 + \frac{1}{2}I_y\omega_y^2 + \frac{1}{2}I_2\omega_z^2$$

Here the total independent terms required is six.

So the degree of freedom of a polyatomic molecule will be six.

2.16 Internal Energy of an Ideal Gas

Monoatomic gases like He, Ne etc. only possess the translational motion i.e. they possess only linear motion.

But if we consider diatomic or polyatomic gases like O_2, H_2, CO_2, NH_3, CH_4 etc. their molecules possess both translational and rotational energy. So the internal energy of the gas is shared between translational and rotational motion of molecules.

The internal energy of an ideal gas does not depend on its volume. It depends only on its temperature. The internal energy associated with a monoatomic gas is

$$\text{Internal Energy (U)} = n\,N\left(\frac{3}{2}k\text{T}\right) = n\cdot\frac{3}{2}\text{RT}\left[\because k = \frac{R}{N}\right]$$

Similarly internal energy associated with a diatomic molecule is

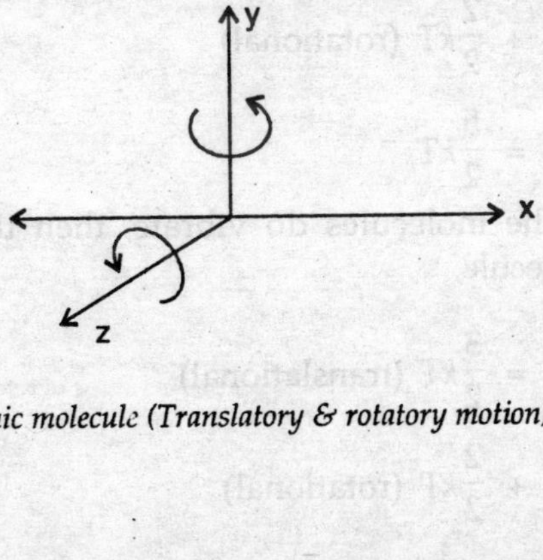

Fig. 2.4. Diatomic molecule (Translatory & rotatory motion).

$$U = n\,N\left(\frac{5}{2}kT\right)$$

$$= n.\ \frac{5}{2}RT$$

where n = amount of gas in moles
N = Avagadro's number
R = Universal gas constant

Average energy per molecule of

a monoatomic gas = $\frac{3}{2}kT$ (Translational)

Average energy per molecule of a diatomic gas (if the molecules do not vibrate)

$$= \frac{3}{2}kT \text{ (translational)}$$

$$+ \frac{2}{2}kT \text{ (rotational)}$$

$$= \frac{5}{2}kT$$

If the molecules do vibrate, then the average energy per molecule

$$= \frac{3}{2}kT \text{ (translational)}$$

$$+ \frac{2}{2}kT \text{ (rotational)}$$

$$= \frac{5}{2}kT$$

If the molecules do vibrate, then the average energy per molecule

$$= \frac{3}{2}kT \text{ (translational)}$$

$$+ \frac{2}{2}kT \text{ (rotational)}$$

$$+ \frac{2}{2}kT \text{ (vibrational)}$$

$$= \frac{7}{2}kT.$$

NUMERICALS

Example 1. *Calculate the RMS velocity of oxygen molecule at 1 atm and 0°C (the density of oxygen is 1.42 g per litre).*

Solution. $C = \sqrt{\dfrac{3p}{d}}$

Given $P = 1 \text{ atom} = 76 \text{ cm of Hg}$

$= 76 \times 13.6 \times 981 \text{ dynes/cm}^2$

$d = 1.429 \text{ g/1} = 0.001429 \text{ g/cc}.$

Substituting these values, we get

$$C = \sqrt{\frac{3p}{d}} = \sqrt{\frac{3 \times 76 \times 13.6 \times 981}{0.001429}}$$

$$= 46120 \text{ cm/s}.$$

Example 2. *Calculate the r.m.s. and average velocity of hydrogen molecules at 0°C.*

Solution. For 1 mole of a gas, we have

$PV = RT = \frac{1}{3} m\, Nc^2$ (N= Avagadro's number)

$= \frac{1}{3} Mc^3$ (M = Mol. wt)

or $C = \sqrt{\dfrac{3RT}{M}}$

Substituting the given values, we get

$$C = \sqrt{\frac{3 \times 8.314 \times 10^7 \times 273}{2.016}} \qquad [\because R = 8.31 \times 10^7]$$

$= 18.4 \times 10^4 \text{ cm/s}.$

Example 3. *Calculate the K.E. in Joules of 1 mole of CO_2 at 27°C, assuming ideal behaviour. What is the average kinetic energy of a single molecule?*

Solution. $E = \frac{3}{2}RT$

$= 3/2 \times 8.314 \times 10^7 \times 300$

$= 3741 \times 10^7$ ergs.

$= 3741$ J.

Average K.E. of a molecule $= \frac{3}{2}\frac{RT}{N_0}$

$= \frac{3741 \times 10^7}{6.02 \times 10^{23}}$

$= 6.214 \times 10^{-14}$ ergs.

Example 4. *Calculate the total kinetic energy, in Joules, of the molecules in 8g of methane at 27°C.*

Solution. K.E. $= \frac{3}{2}RT = 3741$ J (See previous example)

Total energy of 8g of methane $= \frac{1}{2}$ mole of methane

$= \frac{1}{2} \times 3741$

$= 1870.5$ J.

(*Note:* Average K.E. per molecule will be same at the same temperature, irrespective of the nature of the gas).

Example 5. *Calculate in ergs the K.E. of a molecule of monoatomic gas moving with a velocity of 5×10^4 cm sec^{-1}. Given the molecular weight as 20.2.*

Solution. Weight of 1 molecule of gas $= \frac{20.2}{6.02 \times 10^{23}}$

$= 3.355 \times 10^{-23}$ g/molecule

$$\text{K.E. of a single molecule} = \frac{1}{2}mc^2$$
$$= \frac{1}{2}\times 3.355\times 10^{-23}\times(5\times 10^4)^2$$
$$= 4.194\times 10^{-14} \text{ ergs.}$$

2.17 Transport Phenomenon

Though for a gas its equilibrium state is most stable, yet if the equilibrium has not been attained, the following three cases may arise:

(i) The velocities of different parts of gas may be different and so there will be a relative motion of gaseous molecules with respect to one another. As a result of such a motion the faster moving particles will impart momentum to those parts which are moving slowly till an equilibrium is attained. This gives rise to the phenomenon of viscosity.

(ii) The temperature difference may exist in different parts of the gas. In such a situation the K.E. is carried by the molecules of the gas from the region of high temperature to the region of low temperature till the equilibrium is attained. Due to this the phenomenon of conduction arises.

(iii) The molecular concentrations may be different in different parts of the gas. In such a situation the gaseous molecules will carry mass from regions of higher concentration to regions of lower concentration attain the state of equilibrium. This gives rise to phenomenon of diffusion.

These phenomenon which represent the transport of momentum (viscosity), transport of energy (conduction) and transport of mass (diffusion) are known as transport phenomenon.

2.18 Viscosity

Imagine a certain mass of a gas moving in parallel layers between two horizontal planes PQ and RS. If it is assumed that the velocity of gas in contact with layer RS is zero and its velocity goes an increasing as we move towards the plane PQ. Now consider same intermediate plane XY. It can be easily visualised that the layers of the gas above XY are moving faster than the layers of the gas below XY. The faster moving layers above XY exert a tangential force on this plane *xy* and it tend to increase the velocity but simultaneously the slow moving layers below plane XY exert an equal force in the opposite direction. Because of these two opposing forces above and below the plane *xy*, a tangential force is set up, in the plane *xy*, having a tendency to reduce the difference in velocities of the layers on the two sides of the plane *xy*. This property is known as viscosity.

We also find that, due to greater average horizontal velocities, the layers above plane XY has greater momentum in horizontal direction that the layers below XY. Because of thermal agitation same molecules of gas move downwards across XY where as some molecules having smaller average momentum will be moving upwards. Since no transference of mass occurs perpendicular to XY, so the number of molecules moving downwards must be equal to the number of molecules moving upwards. Due to this the slow moving layers below XY gain momentum where as faster moving layers above XY lose momentum. We may thus consider viscosity as a phenomenon of transport of momentum.

If n is the number of molecules per unit volume of a gas and v is their average speed. Since the molecules are moving, because of thermal agitation, in all possible directions so, it may assumed that one-third of the molecules are moving in each of three directions parallel to three co-ordinate axis. So that on an average, one-sixth of the molecules move parallel to any one axis in one particular direction. Therefore, the number of molecules crossing the plane XY upwards or downwards per unit area per second = $nv/6$.

If G is the momentum of each molecule in the plane XY and dG/dz the momentum gradient (rate of change of momentum with distance) in an upward direction perpendicular to XY. If each of the planes RS and PQ are at a distance l from XY where λ is the mean free path.

Then the momentum of each molecules at the plane PQ

$$= G + \lambda \frac{dG}{dz}$$

and the momentum of each molecule at the plane RS = $G - \lambda \frac{dG}{dz}$.

Hence the momentum carried downwards by the molecules crossing unit area of the plane XY per sec. from the plane

$$PQ = \frac{1}{6} n\bar{v}\left(G + \lambda \frac{dG}{dz}\right)$$

Similarly the momentum carried upwards by the molecules crossing unit area of the plane XY per sec. from the plane RS = $\frac{1}{6} n\bar{v}\left(G - \lambda \frac{dG}{dz}\right)$. Hence the net momentum transferred downward per unit area of the plane XY per second.

$$= \frac{1}{6} n\bar{v}\left\{G + \lambda \frac{dG}{dz}\right\} - \frac{1}{6} nv\left\{G - \lambda \frac{dG}{dz}\right\}$$

$$= \frac{1}{3} \bar{v}\lambda \frac{dG}{dz}$$

$$= \frac{1}{3} n\bar{v}\lambda \frac{dG}{dz}(mu)$$

(Since momentum G = *mu*, *u* being the velocity of the molecules at plane XY)

$$= \frac{1}{3} mn\bar{v}\lambda \frac{du}{dz}$$

According to Newton's law, the force is defined as the rate of change of momentum. Therefore, the above expression is a measure of force exerted on the plane XY by the upper layers. Since we are considering unit area, this force is the same as the tangential stress.

The coefficient of viscosity is defined as the tangential force per unit area between two layers needed to maintain a unit velocity gradient in it. i.e.,

$$\text{coefficient of viscosity } \eta = \frac{\text{tangential stress}}{\text{velocity gradient}}$$

$$= \frac{\frac{1}{3} mn\bar{v}\lambda \left(\frac{du}{dz}\right)}{du/z} = \frac{1}{3} mn\bar{v}\lambda \qquad \ldots(1)$$

But mean free path, $\lambda = \frac{1}{\sqrt{2}\pi\sigma^2 n}$

$$\therefore \qquad \eta = \frac{m\bar{v}}{3\sqrt{2}(\pi\sigma^2)} \qquad \ldots(2)$$

Since the above expression does not contain n, coefficient of viscosity is independent of pressure. As the expression contains $\bar{v}$ which is proportional to $\sqrt{T}$; the coefficient of viscosity of a gas varies directly as the square root of the absolute temperature of the gas.

Ex. *If* $\eta = 166 \times 10^{-6}$ *dynes per sq. cm. per unit velocity gradient* $\bar{v} = 4.5 \times 10^4$ *cm/sec., r* $= 1.25 \times 10^{-3}$ *gm/c.c. and n* $= 2.7 \times 10^{19}$ *molecules per c.c. for nitrogen, calculate the mean free path, frequency of collision and the molecular diameter of nitrogen molecules.*

Solution. The coefficient of viscosity is given by

$$\eta = \frac{1}{3} mn\bar{v}\lambda$$

∴ The mean free path

$$\lambda = \frac{3\eta}{mn\bar{v}} = \frac{3\eta}{\rho\bar{v}}$$

Given $\eta = 166 \times 10^{-6}$ dynes per sq. cm. per unit velocity gradient.

$$mn = \rho = 1.25 \times 10^{-3} \text{ gm./c.c.}$$
$$\bar{v} = 4.5 \times 10^{4} \text{ cm./sec}$$

$$\therefore \quad \lambda = \frac{3 \times 166 \times 10^{-6}}{1.25 \times 10^{-3} \times 4.5 \times 10^{4}}$$
$$= 8.8 \times 10^{-8} \text{ cm.}$$

The mean free path is the average distance travelled by the molecule between two successive collisions. Therefore, the average time between two collisions = $\lambda\sqrt{v}$, $\bar{v}$ is average velocity. Thus the number of collisions per sec., i.e., the frequency of collisions.

$$= \frac{\bar{v}}{\lambda}$$

$$= \frac{4.5 \times 10^{4}}{8.8 \times 10^{-6}}$$
$$= 5 \times 10^{9}$$

Also

$$\lambda = \frac{1}{\sqrt{2}(\pi n \sigma^2)}$$

or

$$\sigma^2 = \frac{1}{\sqrt{2}(\pi n \lambda)}$$

$$= \frac{1}{4.414 \times 3.14 \times 2.7 \times 10^{19} \times 8.8 \times 10^{-6}}$$

∴ The molecular diameter

$$\sigma = \frac{1}{\sqrt{(4.414 \times 3.14 \times 2.7 \times 10^{19} \times 8.8 \times 10^{-6})}}$$
$$= 3.66 \times 10^{-8} \text{ cm.}$$

2.19 Thermal Conduction (Transport of Energy)

This phenomenon can be explained just in the manner we have explained the phenomenon of viscosity. Again consider some mass of the gas at rest and the planes RS, XY and PQ are such that the temperature increases as we move from RS to PQ through XY.

Thus the gaseous molecules above XY plane have higher average temperature than the average temperature of gaseous molecules below the plane XY. Because of this temperature difference the gaseous molecules coming from PQ and passing downwards across XY will have higher K.E. of thermal agitation as compared to that of the molecules coming from RS and passing the plane XY upwards. Due to this transference of molecules across XY, there is a continuous increase in average energy of molecules below XY plane and a continuous decrease in average energy of molecules above XY plane. Thus the heat is conducted from PQ to RS till an equilibrium is attained. Hence the phenomenon of thermal conduction may be considered as transport of energy.

If n is the number of molecules per unit volume of a gas and v is their average velocity. Since the molecules are moving, due to thermal agitation, in all possible directions so it may be assumed that are third of the molecules are moving in each of the three directions parallel to three coordinate axis. So that on an average, one-sixth of the molecules move parallel to any one axis in one particular direction. Therefore, the number of molecules crossing the XY plane upwards or downwards per unit area per second is given by $\frac{1}{3}nv$.

If E is the energy of each molecule in the plane XY and dE/dz the energy gradient (rate of change of energy with distance) in an upward direction perpendicular to XY. Let each of the plane RS and PQ be at a distance 1 from XY, where 1 is the mean free path.

The energy of each molecule at the plane PQ = E + $\lambda \frac{dE}{dz}$

and the energy of each molecule at the plane RS = E − $\lambda \frac{dE}{dz}$.

Thus, the energy carried downwards by the molecules crossing unit area of the plane XY per unit time from the plane

$$PQ = \frac{1}{6} \cdot n\bar{v}\left(E - \lambda \frac{dE}{dz}\right)$$

Similarly the energy carried upwards by the molecules crossing unit area of the plane XY per sec from the plane RS.

$$= \frac{1}{6} \cdot n\bar{v}\left(E - \lambda \frac{dE}{dz}\right).$$

Hence the net energy transferred downward per unit area of the plane XY per sec.

$$= \frac{1}{6} \cdot n\bar{v}\left[E - \lambda \frac{dE}{dz}\right] - \frac{1}{6} \cdot n\bar{v}\left[E - \lambda \frac{dE}{dz}\right]$$

$$= \frac{1}{3} \cdot n\bar{v}\lambda \frac{dE}{dz}$$

Since E = heat energy of a molecule = mc_v T where c_v is the specific heat (Joule/kg °C) of the gas constant volume and T is absolute temperature.

Thus the net energy transferred downwards across unit area of the plane per sec.

$$= \frac{1}{3} \cdot n\bar{v}\lambda \frac{d}{dz}(mc_v \mathrm{T})$$

$$= \frac{1}{3} \cdot mn\bar{v}\lambda c_v \frac{d\mathrm{T}}{dz}$$

The coefficient of thermal conductivity (K) of a gas is defined as the ratio of quantity of heat transferred across unit area in one second to the temperature gradient i.e.,

$$K = \frac{\frac{1}{3}mn\bar{v}\lambda c_v \frac{dT}{dz}}{\frac{dT}{dz}}$$

$$= \frac{1}{3}mn\bar{v}\lambda c_v. \qquad ...(1)$$

Since mean free path $\lambda = \frac{1}{\sqrt{2}(n\pi\sigma^2)}$

$$\therefore \quad K = \frac{1}{3\sqrt{2}} \cdot \frac{m\bar{v}c_v}{\pi\sigma^2} \qquad ...(2)$$

From expression (2) it becomes clear that the coefficient of thermal conductivity is independent of pressure. As the expression contains v that changes as $\sqrt{T}$, the coefficient of thermal conductivity of the gas varies directly as the square root of absolute temperature of the gas.

Comparing equation (1) with $\eta = \frac{1}{3}\bar{v}\lambda$ and remembering that $mn = \rho$, we get

$$\frac{K}{\eta} = c_v \qquad ...(3)$$

η being coefficient of viscosity.

From equation (2), we have

$$K = \frac{1}{3\sqrt{2}} \cdot \frac{\bar{v}}{\pi\sigma^2}(m)c_v$$

$$= \frac{1}{3\sqrt{2}} \cdot \frac{\bar{v}}{\pi\sigma^2}\left(\frac{M}{N}\right)\frac{c_v}{M}$$

where M is the molecular weight, N is Avogadro Number and C_v is the molar specific heat at constant volume

$$\therefore \quad K = \frac{1}{3\sqrt{2}} \cdot \frac{c_v}{N\pi\sigma^2}\bar{v}$$

As

$$\bar{v} = \sqrt{\left(\frac{8kT}{mn}\right)}$$

the coefficient of thermal conductivity is inversely proportional to the square root of the mass of the molecule.

Since the mass of hydrogen molecule is least, the thermal conductivity of hydrogen gas will be large as compared to any other gas at any given temperature.

Ex. 28. *The coefficient of viscosity of helium at 273 K and one atmosphere is h_1. that of argon is h_2. The atomic masses of these monoatomic gases are μ_1 and μ_2 respectively. Calculate in terms of the above constants:*

(i) the ratio $\frac{S_2}{S_1}$, where S_2 is the total scattering cross-section of the $A_r - A_r$ atom while S_1 is the total scattering cross-section of the He – He atom.

(ii) the ratio $\frac{K_2}{K_1}$, where K_2 is the thermal conductivity of Argon gas while K_1 is the thermal conductivity of the helium gas when T = 273 K.

Solution. (i) The coefficient of viscosity is given by

$$\eta = \frac{1}{3\sqrt{2}} \frac{m\bar{v}}{3} = \frac{1}{3\sqrt{2}} \frac{m\bar{v}}{S}$$

where S is the scattering cross-section.

For argon, $$\eta_1 = \frac{1}{3\sqrt{2}} \frac{m_1\bar{v}_1}{S_1}$$

For helium $$\eta_2 = \frac{1}{3\sqrt{2}} \frac{m_2\bar{v}_2}{S_2}$$

$$\therefore \qquad \frac{\eta_1}{\eta_2} = \frac{m_1}{m_2} \cdot \frac{\bar{v}_1}{\bar{v}_2} \cdot \frac{S_2}{S_1}$$

$$\therefore \quad \frac{S_2}{S_1} = \frac{\eta_1}{\eta_2} \cdot \frac{m_2}{m_1} \cdot \frac{\bar{v}_2}{v_1} \quad \ldots(1)$$

But

$$\bar{v}_1 = \sqrt{\left(\frac{8kt}{m_1\pi}\right)} \text{ and } \bar{v}_2 \sqrt{\left(\frac{8kt}{m_2\pi}\right)}$$

So that

$$\frac{\bar{v}_2}{\bar{v}_1} = \sqrt{\left(\frac{m_2}{m_1}\right)}$$

$\therefore$ From equation (1),

$$\frac{S_2}{S_1} = \frac{\eta_1}{\eta_2} \cdot \frac{m_2}{m_1} \sqrt{\left(\frac{m_2}{m_1}\right)} = \frac{\eta_1}{\eta_2} \cdot \left(\frac{m_2}{m_1}\right)^{3/2} \quad \ldots(2)$$

We also know $m_2 = \frac{\mu_2}{N}$ and $m_1 = \frac{\mu_1}{N}$
where N is Avogadro number.

$$\therefore \quad \frac{m_2}{m_1} = \frac{\mu_2}{\mu_1}$$

So that equation (2) gives

$$\frac{S_2}{S_1} = \frac{n_1}{n_2} \cdot \left(\frac{\mu_2}{\mu_1}\right)^{3/2} \quad \ldots(3)$$

(ii) The thermal conductivity is given by

$$K = \frac{1}{3\sqrt{2}} \cdot \frac{C_v}{N \cdot \pi\sigma^2} \bar{\upsilon} = \frac{1}{3\sqrt{2}} \cdot \frac{C_v}{NS} \bar{\upsilon}$$

where C_v is molar or atomic specific heat and S is scattering cross-section.

As

$$\bar{v} = \sqrt{\left(\frac{8kT}{m\pi}\right)} = \left(\frac{8kt}{(\mu / N) \cdot \pi}\right) = \sqrt{\left(\frac{8kTN}{\mu\pi}\right)}$$

$$\therefore \quad K = \frac{1}{3\sqrt{2}} \frac{C_v}{NS} \sqrt{\left(\frac{8kTN}{\mu\pi}\right)} = \frac{1}{3\sqrt{2}} \frac{C_v}{NS} \left(\frac{8kt}{\mu N\pi}\right)$$

For Agron gas $K_1 = \frac{1}{3\sqrt{2}} \frac{C_v}{S_2} \sqrt{\left(\frac{8kT}{\mu_1 N\pi}\right)}$

For helium gas

$$K_2 = \frac{1}{3\sqrt{2}} \frac{C_v}{S_2} \sqrt{\left(\frac{8kT}{\mu_2 N\pi}\right)}$$

$$\therefore \quad \frac{K_2}{K_1} = \frac{S_1}{S_2} \sqrt{\left(\frac{\mu_1}{\mu_2}\right)} \qquad \ldots(4)$$

From eqn (3).

$$\frac{S_1}{S_2} = \frac{\eta_2}{\eta_1} \left(\frac{\mu_1}{\mu_2}\right)^{3/2}$$

$$\therefore \quad \frac{K_2}{K_1} = \frac{\eta_2}{\eta_1} \left(\frac{\mu_1}{\mu_2}\right)^{3/2} \sqrt{\left(\frac{\mu_1}{\mu_2}\right)}$$

$$\text{or} \quad \frac{K_2}{K_1} = \frac{\eta_1}{\eta_2} \left(\frac{\mu_1}{\mu_2}\right)^2 \qquad \ldots(5)$$

2.20 Diffusion (Transport of Mass)

Again consider a mass of a gas moving in parallel layers between the planes RS and RQ. XY is an intermediate plane such that the concentration (i.e. number of molecules per unit volume) increases in the vertical direction in going from plane RS to PQ through XY. Thus the concentration of the gas is greater above XY plane then its concentration below this plane. To attain equilibrium, the gaseous molecules move from plane PQ to RS crossing XY and also from RS to PQ crossing plane XY. Thus movement of molecules due to thermal agitation continues till the equilibrium is attained. Thus this phenomenon of diffusion may be thought of *a* the transport of mass.

Since the molecules are moving in all possible directions because of thermal agitation, it may be assumed that one-third of the molecules are moving in each of the three directions parallel to cartesian coordinate axis. Thus on the average one-sixth of the molecules move parallel to any one in one particular direction.

Let n be the concentration at the plane XY and dn/dz the rate of change of concentration with distance in an upward direction perpendicular to XY. Let each of the planes RS and PQ be at a distance λ from XY, where λ is the mean free path. Then concentration at the plane CD

$$= n + \lambda \frac{dn}{dz}$$

And the concentration at the plane AB $= n - \lambda \dfrac{dn}{dz}$

The number of molecules from the plane PQ and crossing the plane XY downwards per unit area per second

$$= \frac{1}{6}\bar{v}\left(n + \lambda \frac{dn}{dz}\right)$$

and the number of molecules coming from the plane RS and crossing unit area of the plane XY per second

$$= \frac{1}{6}\bar{v}\left(n - \lambda \frac{dn}{dz}\right)$$

∴ The net number of molecules crossing unit area of the plane XY downwards per sec.

$$= \frac{1}{6}\bar{v}\left(n + \lambda \frac{dn}{dz}\right) - \frac{1}{6}\bar{v}\left(n - \lambda \frac{dn}{dz}\right) = \frac{1}{3}\bar{v}\lambda \frac{dn}{dz}$$

The coefficient of diffusion is defined as the ratio of the number of molecules across unit area in one second to the concentration gradient i.e.,

$$\text{The coefficient of diffusion } D = \frac{\frac{1}{3}\bar{v}\lambda \frac{dn}{dz}}{\frac{dn}{dz}}$$

$$= \frac{1}{3}\bar{v}\lambda$$

$$= \frac{1}{3}\frac{\rho\bar{v}\lambda}{\rho} \qquad \text{...(1)}$$

$$= \frac{\eta}{\rho} \qquad \ldots(2)$$

where $\eta = \frac{1}{3}\rho\bar{v}\lambda$ = coefficient of viscosity.

But $\qquad \lambda \propto \frac{1}{n} \propto \frac{T}{P}$ and $\bar{v} \propto \sqrt{T}$

$\therefore \qquad D \propto p^{-1}T^{3/2}$

i.e. the coefficient of diffusion is directly proportional to $T^{3/2}$ where T is absolute temperature and is inversely proportional to pressure p.

2.21 Brownian Motion

The random and zig-zag motion of fluid (i.e. gaseous) particles is known as Brownian motion. This phenomenon was first observed by Brown (1827) during microscopic examination of the suspension of five in animate spores in water. The phenomenon can also be observed microscopically in colloidal solutions.

The particle size in colloidal solutions is of the order of 10^{-4} cm, which is about $\frac{1}{10{,}000}$th of the size of molecules. The phenomenon of Brownian motion provides a useful picture of gases state of matter.

It can be thought (though it can not be observed) that gas molecules are in constant random motion in all possible directions. They move in straight lines till they collide with each other or with the walls of the container. All the laws of Brownian particles are applicable in case of gaseous molecules.

Einsteins Theory of the Translational Brownian Motion

This theory tries to explain the physical nature of the phenomenon. Since the Brownian particles tend to diffuse with the median so they can be related to diffusion coefficient of Brownian Motion.

Einestein calculated the diffusion coefficient in following two ways:

(i) from the irregular motion of suspended particles

(ii) from the difference in osmotic pressure between different parts due to difference in the concentration of suspended particles, which in the cause of diffusion in gases:

(i) Calculation of Coefficient of Diffusion from Random Molecular Motion

Consider a system in which molecular concentration is not uniform and the diffusion is caused due to its gradient along x-axis. Suppose that every particle suffers a displacement in Δ time τ.

Consider a cylinder having a cross-sectional area 'A' and having its axis parallel to x-axis. Let the end faces, S_1 and S_2 of the cylinder be separated by a distance Δ. Also let the molecular concentration at the ends S_1 and S_2 of the cylinder be n_1 and n_2 respectively.

Now the number of particles that cross the end face S_2 to the right in time $\tau = \frac{1}{2} \times n_1 \times \Delta A$ [The above relation is valid because half of the particles contained in the cylinder of volume Δ. A situated to left of S_1 will cross it in time τ.]

Similarly the number of particles crossing the end face S_2 to the left in time τ,

$$= \tfrac{1}{2}\, n_2 . \Delta A$$

Hence the excess of particles crossing a middle layer to the right

$$= \tfrac{1}{2}\,(n_1 - n_2)\, \Delta A$$

If dn/dx is the gradient of molecular concentration, then according to definition coefficient, we must have

$$\frac{1}{2}(n_1 - n_2)\Delta A = -\Delta \frac{dn}{dx} \tau A$$

But

$$n_1 - n_2 = -\Delta \frac{dn}{dx}$$

$$\therefore \qquad \frac{1}{2}\cdot\left(-\Delta\frac{dn}{dx}\right) = -\Delta\frac{dn}{dx}\tau A$$

$$\therefore \qquad D = \frac{\Delta^2}{2\tau} \qquad \ldots(1)$$

(ii) Calculation of coefficient of diffusion from the difference in osmotic pressure between different parts due to differences in concentration of the suspended particles. Let the osmotic pressure acting on the end faces S_1 and S_2 of the cylinder be p_1 and p_2. From the gas laws $p_1 = n_1\,kT$ and $P_2 = n_2kT$. Due to this difference of pressure, the cylinder will experience a force equal to

$$(p_1 - p_2)\,A = (n_1kT - n_2kT)\,A$$

$$= (n_1 - n_2)\,kTA.$$

This force pushes the cylinder in the positive direction of X-axis.

If n is the mean concentration, then this force acts on $n.A\Delta$ particles contained in the cylinder. Thus the force acting on a single particle is given by

$$f = \frac{(n_1 - n_2)kTA}{nA\Delta}$$

$$= \frac{kT}{n}\cdot\frac{n_1 - n_2}{\Delta} = \frac{kT}{n}\left(-\frac{dn}{dx}\right)$$

$$= \frac{kT}{n}\cdot\frac{dn}{dx}$$

Thus according a Stoke's law the particle being suspended in viscous medium and assumed spherical will have a steady velocity v given by the relation

$$f = 6\pi\eta r v = -\frac{kT}{n}\frac{dn}{dx}$$

where r is radius of the particle and η is the coefficient of viscosity.

$$\therefore \qquad nv = \frac{kT}{6\pi\eta r}\cdot\frac{dn}{dx}$$

This expression represents the number of particles moving to the right per unit area.

According to definition of diffusion coefficient.

$$nv = -D\cdot\frac{dn}{dx}$$

$$\therefore \quad -D\frac{dn}{dx} = \frac{kT}{6\pi\eta r}\cdot\frac{dn}{dx}$$

$$D = \frac{kT}{6\pi\eta r}$$

$$D = \frac{RT}{N}\frac{1}{6\pi\eta r} \quad \ldots(2)$$

$$\left(\text{since } k = \frac{R}{N}\right)$$

where N is Avogadro number.

Equating (1) and (2), we get

$$\Delta^2 = \frac{RT}{N}\frac{1}{3\pi\eta r}t. \quad \ldots(3)$$

Perrin made measurement of all the quantities in eqn. (2) except N and then calculated the value of N. The value of N found in this way was in agreement with the accepted value.

2.22 Molecular Beams

A molecular beam is defined as a narrow stream of molecules in which the individual molecules go their own way and do not interact with others.

The beam of molecules is directed towards other molecules, and the scattering that takes place due to the impact depends upon inter molecular forces. It is possible to control the speed of the molecules in the beam and to the behaviour of molecules as they collide with the target molecule having different kinetic energy can be investigated. The molecular beams can be produced at a pressure before about 10^{-5} Torr.

Application of Molecular Beams

In the recent least a notable advance in the experimental study of intermolecular forces has come from the development of molecular beams and molecular beams are extensively used in the study of molecular properties.

(1) Determination of Intermolecular Potential

Molecular beams have been used to examine intermolecular potential functions.

(2) Verification of the Maxwell's Law of Distribution of Molecular Velocities

The experimental verification of the Maxwell's law of distribution of molecular velocities have been made in molecular beam study by Miller and Kusch (1955).

In Maxwell's law of distribution of molecular velocities, $\frac{dn_c}{n}$, the fraction of the total number of molecules having velocities between C and $(c + dc)$ may the determined and hence the law may be verified.

(3) In the Study of Reaction Mechanism

Molecular beam have been used in the study of reaction mechanisms

$$K + H\,Br \rightarrow H + K\,Br$$

$$K + Br_2 \rightarrow Br + K\,Br$$

$$K + CH_3I \rightarrow CH_3 + K\,I$$

$$D + H_2 \rightarrow H + HD$$

$$Cs + Rb\,Cl \rightarrow Rb + Cs\,Cl$$

$$Cl + Br_2 \rightarrow Br + Br\,Cl$$

(4) In the Study of Long Lived Collisions

Molecular beams have also been used to study long lived collisions

$$Cs\,Cl + K\,I \rightleftharpoons [Cs\,Cl.\,K\,I] \rightleftharpoons Cs\,I + KCl$$

3

Real Gases or Non-Ideal Gases

3.1 Introduction

The experimental results have revealed that the gas laws are only approximately obeyed by the various gases. It is only at very low pressures and relatively high temperatures, the gases tend to behave in accordance with the gas laws. The equation PV = nRT is quite insufficient for accurate measurements. A gas, really a hypothetical one, which follows the gas laws rigorously under all circumstances is called an *ideal gas,* as distinct from real gases. A *real gas* exhibits some deviations from ideal gas laws. When these deviations become appreciable, the gas is said to behave as a real gas or non-ideal gas or imperfect gas.

3.2 Ideal and Real Gases

A gas is termed as ideal if it obeys Boyle's law, Charles law and gas equation (PV + nRT) under all condition of temperature and pressure. However they obey these laws only approximately.

Generally those gases which are not easily liquified and which do not readily dissolve in water show this type of behaviour and only a small deviation from the gas laws. e.g. H_2, N_2 etc.

However those gases which are easily liquified and are readily soluble show much larger deviations, we thus observe large deviations in case of gases like NH_3, CO_2, SO_2 etc. At low pressures and high temperatures these deviations become less pronounced and these gases can be considered to behave as *ideal gases* strictly speaking *no gas is ideal* and *the concept of ideal gas is hypothetical.* Ordinarily the gases are found to obey gas laws fairly well if pressure is low or temperature is high. Such gases are therefore called *Real gases.*

Hence a *real gas* may be defined as *a gas which obeys gas laws fairly well at low pressure or high temperature.*

The higher the pressure and lower the temperature, the greater are the deviations from the ideal behaviour. Generally, the most easily liquifiable and highly soluble gases show larger deviations. Thus highly soluble gases like CO_2, SO_2, and NH_3 show much larger deviations than H_2, O_2, N_2 etc. which are less soluble.

3.3 Deviations of Real Gases from Gas Laws

(a) Deviations from Boyle's Law

To illustrate the deviation of real gases from Boyle's law consider a plot of PV *versus* P at 0°C for a number of gases as shown in Fig. 3.1.

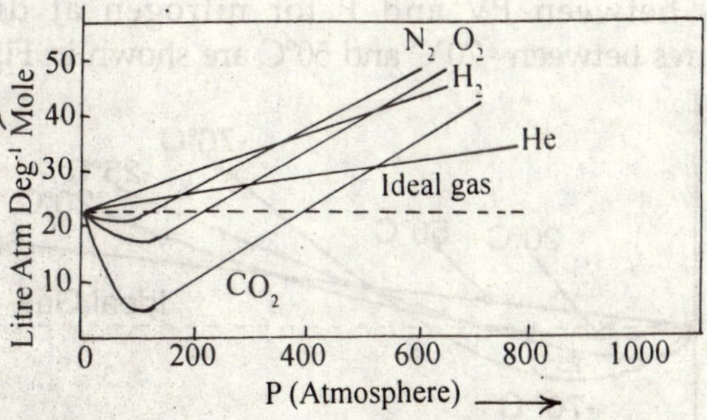

Fig. 3.1. Plot of PV against P for different gases at 0°C.

If Boyle's law is to be obeyed, then at a constant temperature, PV should remain constant at all pressures and the graph should be a horizontal line parallel to the abscissa as shown by dotted line in figure about (Fig. 3.1).

But in actual practice, no gas gives a horizontal straight line, whereas two types of curves are generally obtained which are given below:

(i) In case of some gases like CO_2 (CO, CH_4 etc), the PV decreases with increase of pressure at a constant temperature. This behaviour continues with increase in pressure till PV passes through a minimum at a certain stage. Further increase of pressure also increases PV (Fig. 3.1).

(ii) In case of hydrogen and helium, the product PV continuously goes on increasing with increase of pressure (Fig. 3.1).

However, if the temperature is very much low (–165°C for hydrogen and –240°C for helium), these gases also behave as carbon monoxide, methane etc. at 0°C. Whereas if temperature is quite high, carbon monoxide and methane etc. also behave as hydrogen and helium at 0°C.

(a) Effect of Temperature on Deviations from Ideal Behaviour

Plots between PV and P for nitrogen at different temperatures between –70°C and 50°C are shown in Fig. 3.2.

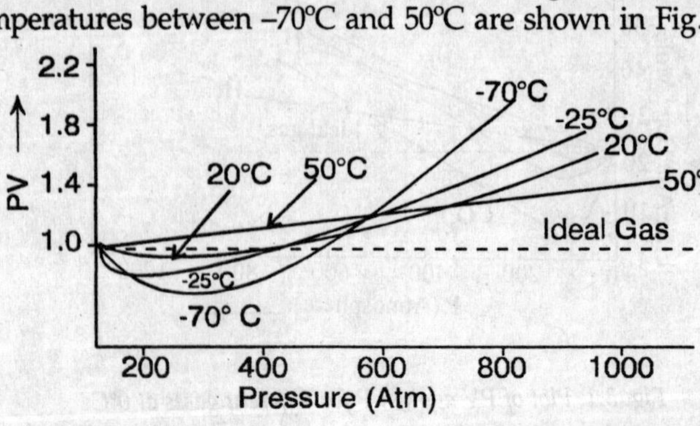

Fig. 3.2. Effect of temperatures on deviation of nitrogen from ideal behaviour.

From Figure 3.2 it is evident that with increase of temperature, the dip in the curve becomes smaller and smaller. At 50°C, it is observed that the curve remains almost horizontal for an appreciable range of pressure lying between about 0°C and about 100 atmospheres. Thus Boyle's law is obeyed with in this range of pressure at 50°C. *This temperature at which a real gas behaves nearly like an ideal gas over an appreciable range of pressure* is called *Boyle Temperature* or *Boyle Point*. Different gases have different Boyle temperatures e.g.

Gas	*Boyle Temperature*
Hydrogen	–165°C
Helium	–240°C

(b) Deviations from Charles—Gay Lussacs Law

According to Charles—Gay Lussacs law, the coefficient of increase in volume for all gases should be 1/273 of its volume at 0°C, at all pressures. But Amagat and others as the basis of their experimental work showed that this is true only at low pressures. As the pressure is increased, considerable deviations set in. In this case also, the gases which are more easily liquifiable show longer deviations.

(c) Deviations from Avogadro's Law

According to Avogadro's law, one mole of any gas should have a volume of 22.4 litres at S.T.P. and it should not depend on the nature of the gas. However, in actual practice, it has been observed that this law is also obeyed by the gases only at low pressures. The volume occupied by 1 mole of some of the gases at N.T.P. is tabulated in Table 3.1 to illustrate it.

Table 3.1
Deviations from Avogadro's Law

Gas	Volume occupied by 1 mole at N.T.P.
Oxygen	22.344 litres
Nitrogen	22.402 litres
Hydrogen	22.427 litres
Carbon dioxide	22.264 litres
Ammonia	22.084 litres

Thus ammonia which is easily liquifiable gas shows maximum deviation.

The non-ideal behaviour of these *real gases* can be described by modifying the ideal gas equation (PV = *n*RT) to

$$PV = nRT\,(1 + bP)$$

A variety of equations such as above, with *b* evaluated for each gas, will be used to describe the complex and highly individualistic behaviour of real gases.

Compressibility Factor Z

Real and Ideal gases can be compared at various pressures and various temperatures by noting the extent to which the value of $\frac{PV}{RT}$ deviates from 1. The quantity $\frac{PV}{RT}$ is given by the symbol Z and is called *compressibility Factor* i.e.

$$Z = \frac{PV}{RT}$$

where V is the volume occupied by 1 mole of the gas at a given temperature and pressure.

For an ideal gas, Z = 1 and is independent of pressure and temperature. For a real gas Z is function of both temperature and pressure and its volume varies. The amount by which actual factor differs from a value of unity is the measure of deviation from the ideal behaviour.

Now $\quad (PV)_{ideal} = RT$

$$\therefore \quad Z = \frac{PV}{RT} = \frac{PV}{(PV)_{Ideal}}$$

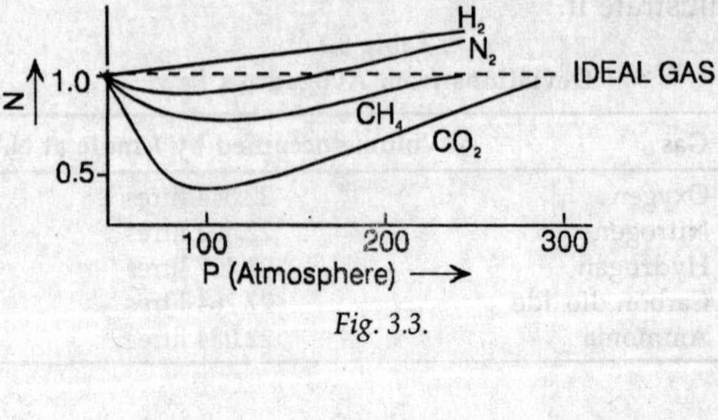

Fig. 3.3.

In other words, the compressibility factor is an index of deviation from ideality.

A plot of Z versus P for some gases at 0°C is shown in Fig. 3.3.

It can be observed from Fig. 1.2 that for H_2 the value of Z continuously increases but for N_2, CH_4 and CO_2, Z first decreases and then increases quite rapidly with the increase of pressure.

It has been found that for those gases which can be easily liquified, Z decreases sharply below the value of unity at low pressure. As carbon dioxide is a gas which can be more easily liquified than nitrogen and methane the dip in value of Z is much more pronounced in carbon dioxide and persists over a range of pressures. But hydrogen gas cannot be easily liquified, therefore, its curve does not exhibit any dip in value of Z with increasing pressure.

3.4 Causes of Deviation from Ideal Behaviour

The failure of the ideal gas equation PV = *n*RT over an extensive range of temperatures and pressures has been attributed by Vander Waals to be due to the following two *faulty assumptions* made in the *kinetic theory*.

(i) According to one of the postulates of kinetic theory:

The volume occupied by the gaseous molecules themselves is negligibly small as compared to the total volume occupied by the gas.

This postulate holds good only under ordinary conditions of temperature and pressure (i.e. pressure is low and temperature is high). But if the pressure becomes very high, say of the order of 100 atmospheres, the total volume of the gas will decrease very much whereas the volume of the molecules will remain almost sameas the molecules are *incompressible*. Hence *under such conditions, the volume occupied by the gaseous molecules will no longer be negligible as compared to the total volume occupied by the gas.*

The same thing is also observed when the temperature is lowered appreciably.

Thus we conclude that the *above postulate* of the kinetic theory of gases *does not hold good at high pressures and low temperatures.*

(ii) According to another postulate of kinetic theory, the *forces of attraction between gaseous molecules are negligible'.*

This postulate also holds good only at low pressures or at high temperatures.

But as the pressure becomes high or temperature become low, the volume also becomes very small and the molecules come closer to each other, so that the intermolecular forces of attraction become appreciable and can not be ignored.

Here are again conclude that this *postulate* of kinetic theory of gases *does not hold good under conditions of high pressures and low temperatures.*

This means that suitable corrections must be applied to ideal gas equation to make it applicable to real gases.

4

Equations of State for Real Gases

4.1. Introduction

Based on the postulates of kinetic theory of gases, it is possible to deduce an equation for the pressure of an ideal gas. This is known as *kinetic equation for gases.*

$$PV = \frac{1}{3}mne^2$$

P and V are pressure and volume respectively of n molecules of a gas, each having mass m, c is the root mean square velocity of the molecules.

The kinetic gas equation is derived from kinetic theory. However in case of an ideal gas the gas equation called *equation of state* is used. It is

$$PV = RT$$

This equation is based on various gas laws and is thus an experimental formulation. However their equation makes use of two basic assumptions of kinetic theory i.e.

(i) The gaseous molecules are point masses, and

(ii) There are no intermolecular forces between molecules.

The above equation (*i.e.* PV = RT) has been found to be

valid to a large extent at ordinary conditions of temperature and pressure. However at high pressure and low temperature, the real gases show a deviation from ideal behaviour.

Deviations from Ideal Gas Equation

To explain the deviation of real gases from ideal behaviour under the conditions of low temperature and high pressure Vander Waals found fault with two basic postulates of kinetic theory of gases on which the ideal gas equation is based. He observed that

(i) Read gases do have molecular volumes under ordinary conditions, free space between the molecules is much larger in comparison to the molecular volume and therefore, neglecting the volume occupied by these gaseous molecules does not involve any measurable error. But at high pressures and low temperatures, when molecules of gas are relatively close together, molecular volume constitutes a significant fraction of the total volume. Thus the volume of the ideal gas should be lower than that of real gas.

(ii) Moreover under conditions of high pressures and low temperatures, when molecules are brought quite close together, some intermolecular attractions come into play. The pressure exerted by the gas get reduced and therefore, ideal pressure should be higher than the real pressure.

Taking into account these factors Vander Waals suggested a modified form of gas equation for real gases.

4.2 Equations of State for Real Gases

Equation of state: To describe the state of a system, we need certain variables. The state of a system can be completely defined by defining the values of a few of the variables and it is not necessary to define the values of macroscopic properties.

For a gaseous system, the state of the system can be specified by its pressure (P), volume (V) and temperature (T). In practice, it is not necessary to specify all the three thermodynamics variables (i.e. P, V and T), since these properties are interrelated for an ideal gas by PV = RT. The value of any of these can be calculated provided those of the other two are known. Hence only two of these three are independent variables.

The equation concerning the pressure, the volume and the absolute temperature of a substance is called the equation of state of that substance.

For a gaseous system the three variables are connected by a relationship of the general form

$$f(P,V,T) = 0$$

which is called the *equation of state.*

For every thermodyanamics system, we have an equation of state which is independent of others. For determining the equation of state of a system, the thermodynamic variables of the system are measured accurately and then a relationship is expressed between them. Such an equation can be obtained only if a system is in equilibrium and it is applicable only to those systems which are in equilibrium state.

The *equation of state for an ideal gas* is

$$PV = RT$$

Where R is called universal gas constant and its value is 8.314J/g-mole K and V is known as Molar volume (i.e. volume occupied by 1 mole of gas)

The *equation of state for a real gas is*

$$\left(P+\frac{a}{V^2}\right)(V-b) = RT$$

This equation is also known as *Vander Waal's equation.*

In this equation *a, b* are known as *Vander Waals constants.*

Vander Waals Equation

I.D. Vander Waals a Dutch physicist, in 1873 proposed an equation of state. For one mole of a gas it can be written as

$$\left(P+\frac{a}{V^2}\right)(V-b) = Rt$$

'*a*' and '*b*' are Vander Waals constants and theory depend upon the nature of the gas.

'*b*' denotes the volume excluded from molecular motion for each mole of the gas having a molar volume *v*. The increased impact rate of the molecules in the restricted volume ($V - b$) will cause the kinetic pressure of the gas due to kinetic energy to increase to

$$\left(\frac{RT}{V}\right)\left(\frac{V}{V-b}\right) = \frac{PT}{V-b}$$

The cohesive forces between molecules may thus be considered to decrease rapidly with distance between the molecules and so this force is appreciable only at small distances.

A molecule at a large distance from the wall of the container is attracted by its neighbouring molecules from all sides and thus the net effect is zero. However for a molecule near the wall of the container, there is an average net force normal to the wall and this tends to pull the molecule from the wall and thus the pressure on the wall is reduced. The decrease in pressure is proportional to the number of molecules being pulled and is also proportional to the number of molecules that perform the pulling. Thus the cohesive pressure because of potential energy must be proportional to the square of the density of the gas and can be written as $\frac{a}{V^2}$ where '*a*' is a constant.

Thus the net pressure P is given by

$$P = \frac{RT}{V-b} - \frac{a}{V}$$

Rearranging we can write

$$\left(P + \frac{a}{V^2}\right)(V-b) = RT$$

This is Vander Waals equation of state.

The explanations for volume correction and pressure correction are given below:

(a) Volume Correction

The assumption that actual volume of the molecules is negligible does not introduce any appreciable error at low pressures. However at high pressures when the molecules and quite close together (i.e. the total volume of gas is small) the volume occupied by molecules themselves becomes a considerable fraction of the total volume occupied by the gas.

According to Vander Waals equation the actual free space available inside the vessel for the movement of the molecules is not V, the volume of the container, but $(V - b)$ where 'b' is the '*effective volume*' of the molecules. Detailed calculation have shown that the *effective volume* of the molecules in motion is equal to 4 times the actual volume of the molecules at rest. Assuming the molecules to the spherical in shape, each of radius, r, consider two such molecules as shown in Fig. 4.1(*A*). It is clear that around each molecule, the centre of no other molecule can enter into sphere of radius, πr the volume excluded by this pair of molecules,

$$V_{pair} = \frac{4}{3}\pi(2r)^3 = 8\left(\frac{4}{3}\pi r^3\right)$$

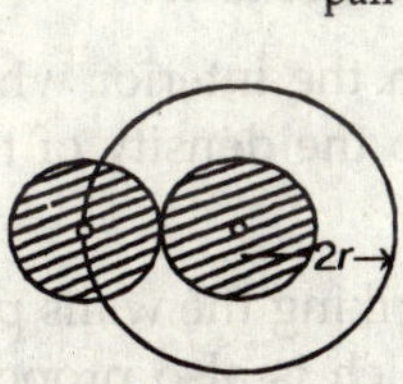

Fig. 4.1 (A)

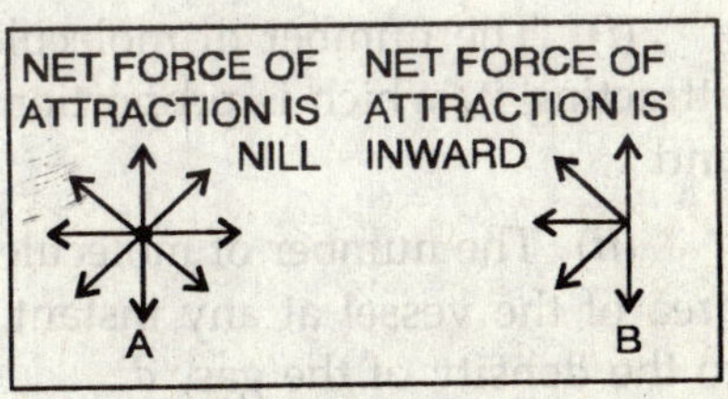

Fig. 4.1 (B)

∴ Volume excluded by single molecules,

$$V_e = \frac{1}{2} \times 8\left(\frac{4}{3}\pi r^3\right) 4V_M$$

where V_M is the actual volume of the molecule. Thus we find that effecting volume V_e of a molecule is four times the actual volume.

2. **Pressure Correction.** This assumption that molecules *exert no appreciable inter-molecular attraction* is not justified especially at high pressure when *the molecules are quite close together*.

Consider particles A and B in the gas. The net force of inter-molecular attraction on particle 'A' situated in the interior of the gas is nil, because it is surrounded by molecules which are equally distributed in opposite directions, thereby the attraction on all sides are naturally cancelled.

However, molecule '*B*' just near the boundary which is about to strike the wall, is subjected to a net *inward force of attraction or pull* due to the interior molecules [Fig. 3.4(B)]. So, when a molecule is about to strike the wall, this inward pull acts so as to reduce the force exerted by the impact and consequently *the measured pressure P is less than the ideal pressure i.e.,* when no attraction were there. This will tend to make PV less than RT and so will tend to cause a negative deviation. Hence a correction term P′ should be added to P to get ideal pressure.

$$\therefore \quad P_{\text{ideal}} = P + P'$$

The inward force, *P′* exerted on molecules about to strike the walls of the vessel is proportional to:

(*i*) The number of molecules in the interior which are attracting it, which is proportional to the density of the gas; and

(*ii*) The number of molecules striking the walls per unit area of the vessel at any instant, which is also proportional to the density of the gas, *d*.

$$\therefore \qquad P' \propto d^2$$

Now density $d \propto \dfrac{1}{\text{Volume}}$

$$P' \propto \frac{1}{V^2}$$

or $$P' = \frac{a}{V^2}$$

Here a is a proportionality constant known as coefficient of attraction.

$$\therefore \qquad P_{\text{ideal}} = P + \frac{a}{V^2}$$

Making both the pressure and volume corrections, the general gas equation for one mole of a gas.

$PV = RT$ may be written as

$$\left(P + \frac{a}{V^2}\right)(V - b) = \text{RT}$$

This equation is known as Vander Waals equation and represents the behaviour of real gases over a wide range of temperature and pressure.

If there are n moles of the gas occupying volume V, then the effective volume of the molecules is given by nb and the actual free space available inside the vessel for the movement of the molecules is then V – nb.

The pressure correction $P' \propto n^2 d^2$

$$\propto a^2 \times \frac{1}{V^2}$$

$$P' = \frac{n^2 a}{V^2}$$

Therefore, the Vander Waals equation for n moles of a gas will be

$$\left(P + \frac{n^2 a}{V^2}\right)(V - nb) = n\text{RT}.$$

4.3 Vander Waals Constants

The units for Vander Waals constants a and b depend on the units in which P and V are expressed.

We are seen that for n molecules

$$P' = \frac{an^2}{V^2}$$

It is clear that the constant a is expressed by pressure × (Volume)^2/mole^2. If pressure is expressed in atmospheres and volume in litres, a will be expressed in atmospheres-litre^2 mole^{-2}.

As b is evidently volume, it is expressed in units of volume, *i.e.*, litre.mole^{-1}.

In S.I. units, the units of

$$a = \frac{\text{Pressure (volume)}^2}{\text{mol}^2}$$

or $$a = \frac{Nm^{-2}(m^3)^2}{\text{mol}^2} = Nm^4\ \text{mol}^{-2}$$

$$\frac{1\ \text{atm litre}^2}{\text{mol}^2} = 1\ \text{atm(dm}^6\text{) mol}^{-2}$$

$$= \frac{1.01325 \times 10^5\, Nm^{-2} \times 10^{-6}\, m^6}{\text{mol}^2}$$

and the units of $$'b' = \text{volume mol}^{-1}$$

$$b = m^3\ \text{mol}^{-1}$$

$$1\ \text{litre mol}^{-1} = dm^3\ \text{mol}^{-1}$$

$$= 10^{-3} \times m^3\ \text{mol}^{-1}.$$

4.4 Vander Waals Equation and Real Gas Behaviour

Vander Waals equation tries to explain the observed deviation for real gases from ideal behaviour. The explanation for high pressure, high temperature and low pressure are as under:–

Vander Waal's equation is

$$\left(P+\frac{a}{V^2}\right)(V-b) \;=\; \text{RT} \qquad \ldots(1)$$

On multiplication and neglecting very small fraction, the equation can be written as

$$PV\text{–}Pb+\frac{a}{V} \;=\; P'V' \qquad \ldots(2)$$

where P′ and V′ is the pressure and volume of the gas if the gas was ideal.

(*i*) **At low pressure.** The term *a*/2 which represents the effect of attraction between the molecules predominates the Pb factor. Hence, neglecting Pb in equation (2) we get

$$PV+\frac{a}{V} \;=\; P'V'$$

or $$PV \;=\; P'V'-\frac{a}{V}$$

It means that the observed product *PV* at low pressure is less than the product of pressure and volume (*P′V′*) if the gas was ideal.

(*ii*) **At high pressure.** The term *Pb* overweights the term $\frac{a}{V}$ and, therefore, the equation (2) can be written as

$$PV-Pb \;=\; P'V'$$

or $$PV \;=\; P'V'+Pb.$$

It means that the observed product *PV* is greater than *P′V′*, the product of pressure and volume of the gas was ideal.

(*iii*) **At high temperature.** *V* is very large, *Pb* and $\frac{a}{V}$ will be negligible small. Hence in this case Vander Waals equation almost becomes $PV = P'V'$ indicating that at high temperature the behaviour of the gas very nearly agrees with Boyle's law.

Exceptional Behaviour of Helium and Hydrogen

As shown in Fig. 4.1, H_2 and He, unlike other gasest show positive deviation from ideal gas behaviour at all pressures. This is because of the fact that both of them have small masses and inter molecular attractions in them are quite low. So the term $\frac{a}{v^2}$ is always negligible in their case. Thus $PV = RT + pb$.

This explains their positive deviation throughout.

Limitations of Vander Waals Equation

This equation can explain the behaviour of gases satisfactorily over a wide range of temperature and pressure. However at very low temperature and very high pressure the results are not in agreement with experimental values. This is because of the fact that constants *a*, *b* are empirical in nature and their values vary with temperature as well as with the nature of the gas. That it is essential to select different values of *a* and *b* over different ranges of temperature and pressure so that the theoretical results are in agreement with the experimental results.

4.5 Other Equations of State

Many more equations have been put forward as a modification of Vander Waals equation. All these are essentially empirical in nature. Some of them are.

(*i*) **Dietericis equation.** Dieterici, took into account the effect of inter molecular forces of attraction on pressure and gave the following equation:

$$P(V - b) = RTe^{-\propto/RTV} \quad \text{...(5)}$$

This equation gives results which are in agreement to experimental values when the pressure is high.

(*ii*) **Clausius equation.** Clausius replaced $\frac{a}{V^2}$ of the Vander Waals equation as he took into account the variation of '*a*' with temperature. He proposed the following equation:

$$\left(P+\frac{a}{T(V+c)^2}\right)(V-b) = RT$$

(*iii*) **Berthelots' equation.** He proposed the following equation:

$$\left(P+\frac{a}{TV^2}\right)(V-b) = RT$$

(*iv*) **Beattie-Bridgeman Equation**

The equation can be written as

$$P = \frac{RT\left(1-\frac{C}{VT^3}\right)}{V^2}\left[V+B_0\left(1-\frac{b}{V}\right)\right]-\frac{A_0}{V^2}\left(1-\frac{a}{V}\right) \quad \ldots(6)$$

In addition to the gas constant R, in this equation, we find five specific constants (A_o, a, B_o, b, c) for a gas. Values of these constants for a large number of gases have been found experimentally.

This equation is reasonably accurate when the volumes involved in calculations are greater than twice the critical volumes. In the critical region it is inaccurate.

(*v*) **Benedict-Webb Rubin Equation**

It is entirely empirical equation and was obtained by putting experimental data of light hydrocarbon. It may be written as

$$P=\frac{RT}{V}+\left(B_0RT-A_0-\frac{C_0}{T^2}\right)\frac{1}{V^2}+(bRT-a)\frac{1}{V^3}+\frac{\alpha}{V^6}$$

$$+\ \frac{C\left(1+\frac{\gamma}{\zeta^2}\right)}{T^2}\cdot\frac{1}{V^3}e^{-\gamma/v^2} \quad \ldots(7)$$

In this equation there are eight parameters B_o, A_o, C_o, *b*, *a*, *c*, α and γ are numerical constants which are different for different gases.

This equation is sufficiently accurate in describing *P–V–T* relation of real gases upto densities of 1.8 times the critical density.

(*vi*) Martin and Hou Equation

$$P = \frac{RT}{V-b} + \frac{A_2 + B_2T + C_2e^{-5.475\frac{T}{T_c}}}{(V-b)^2} + \frac{A_3 + B_3T + C_3e^{-5.475\frac{T}{T_c}}}{(V-b)^3} + \frac{A_4}{(V-b)^4} + \frac{B_5T}{(V-b)^5} \quad \ldots(8)$$

In this equation we have nine specific constants (i.e. *b*, A_2, B_2, C_2, A_3, B_3, C_3, A_4 and B_5).

Later on this equation has been modified by adding to it one more term

$$\frac{A_5 + C_5e^{-5.475\frac{T}{T_c}}}{(V-b)^5}$$

By addition of this terms the number of specific constants has increased to 11 from 9 the new specific constants being A_5 and C_5.

The values of various specific constants have been found for a number of gases.

This equation is valid for both polar and non-polar gases upto densities of 1.5 times the critical density.

In general, the more specific constants used in an equation of state, the more accurate the equation can be. However, it is not always so.

One of the simpler equations that is quite accurate over a considerable wide range has only two specific constants. This equation is Redlich-Kwong Equation.

It can be written as

$$P = \frac{RT}{(V-b)} - \frac{a}{T^{1/2}V(V+b)} \qquad ...(9)$$

It has only two constants (i.e. *a*, *b*). These constants can be evaluated from critical constants. These are

$$a = \frac{0.42748R^2T_c^{2.5}}{P_c}$$

and

$$b = \frac{0.08664RT_c}{P_c}$$

This equation given satisfactory results above the critical temperature for any pressure.

(*vii*) VIRIAL Equation of State

The general equation of state proposed by Kammerling-Onnes in 1901. It is general equation of state. It can be written as

$$PV = A + BP + CP^2 + DP^3 + ... \qquad ...(10)$$

In this equation A must be equal to RT to satisfy the ideal gas equation at zero density.

So $$\frac{PV}{RT} = 1 + \frac{B}{V} + \frac{C}{V^2} + \frac{D}{V^3} ...$$

In this B, C, D etc. are called second third, fourth etc. One called *virial coefficients*.

These coefficients are functions of temperature and also of gas.

The virial equation can also be expressed in terms of a power series of pressure as

$$\frac{PV}{RT} = 1 + B'P + C'P + D'P^3 + ...$$

where the coefficients B', C', D' etc. are functions of temperature of the gas.

Any of the equations of state can be expanded in the virial form e.g. we can expand the Vander Waals equation in the virial form as follows:

$$PV = RT\left(1-\frac{b}{V}\right)^{-1} - \frac{a}{V} \qquad \text{[Vander Waals Equ.]}$$

using Binomial theorem, we get

$$PV = RT\left(1+\frac{b}{V}+\frac{b^2}{V^2}+\frac{b^3}{V^3}+\ldots\right)-\frac{a}{V}$$

or

$$PV = 1+\frac{b-\left(\dfrac{a}{RT}\right)}{V}+\frac{b^2}{V^2}+\frac{b^3}{V^3}+\ldots$$

This is the Vander Waals equation in the virial form in terms of density. The virial coefficients are

$$\mathrm{B} = b-\left(\frac{a}{RT}\right); C=b^2; D=b^3 \text{ etc.}$$

Boyles' Temperatures (T_B): The temperature at which a gas shows ideal gas behaviour over a considerable range of pressure is known as Boyles temperature.

Relation between Boyles temperature (T_B) and Vander Waals Constants (a, b)

$$T_b = \frac{a}{b\cdot R}.$$

4.6 Equation of State and Second Virial Coefficient

We can write the Vander Waals equation as

$$\mathrm{PV} = \frac{RT}{\left(1+\dfrac{a}{PV^2}\right)\left(1-\dfrac{b}{V}\right)}$$

The correction terms $\left(i.e. \dfrac{a}{PV^2} \text{ and } \dfrac{b}{V}\right)$ are both small, of the gas is not too much compressed these terms can be neglected.

Using the binomial theorem and neglecting squares and products of correction terms, the equation becomes

$$PV = RT\left(1+\frac{b}{V}-\frac{a}{PV^2}\right)$$

$$= RT+\frac{RT}{V}b-\frac{RT}{PV^2}\cdot a$$

The second and third terms of R.H.S. are small correction terms. Thus by approximation, we can write the equation, using the approximation as $PV = RT$ is above equation, it becomes

$$PV = RT+\left(b-\frac{a}{RT}\right)P$$

According to Vander Waals equation the second virial coefficient B is given by

$$B = b-\frac{a}{RT}$$

The *Boyles temperature* (T_B) is given by

$$B = 0$$

Thus $$T_B = \frac{a}{Rb}$$

We also known, $T_C = \dfrac{8a}{27R\cdot b}$ [T_c = critical temp.]

Thus according to Vander Waals equation,

$$\frac{T_B}{T_C} = \frac{27}{8} = 3.375$$

If we now consider Bertholet equation, which may be written as

$$PV = RT+\frac{9}{128}\cdot\frac{RT_c}{P_c}\left(1-\frac{8T_c^2}{T^2}\right)P$$

According to this equation the second virial coefficient B is given by

$$B = \frac{9}{128} \cdot \frac{RT_c}{P_c}\left(1 - \frac{6T_c^2}{T^2}\right)$$

and the Boyles Temperature is given by

$$\frac{T_B}{T_C} = \sqrt{6} = 2.45$$

This value is found to be in better agreement with experimental results as compared to the ratio of $\frac{T_B}{T_C}$ predicted by Vander Waals equation.

4.7 Principle of Corresponding States

We can define reduced temperature (θ), reduced pressure (π) and reduced volume (ϕ) as

$$\theta = \frac{T}{T_c}$$

$$\pi = \frac{P}{P_c}$$

$$\phi = \frac{V}{V_c}$$

where T_c, P_c and V_c are critical temperature, critical pressure and critical volume.

Thus θ is the ratio of the temperature of the gas to its critical temperature, π is the ratio of pressure of gas to its critical pressure and ϕ is the ratio of volume of gas to its critical volume.

From above ratios, we get

$$T = \theta \, . \, T_c$$
$$P = \pi \, . \, P_c$$
$$V = \phi \, . \, V_c$$

Substituting these values of P, T, V, and also the values of T_c, P_c, V_c in the Vander Waals equation, we get

$$\left(\pi + \frac{3}{\phi^2}\right)(3\phi - 1) = 8\theta \qquad ...(i)$$

Equation (i) is called the *reduced equation of state* of Vander Waals.

In this equation we find that there are no constants that depend on the nature of the substance under consideration. Thus if for any two substances any two of the quantities θ, π and ϕ are same, then the third will also be same. This is called *law of corresponding states.*

This principle of corresponding states is valid, if

(i) The effect of quantization of the translational degrees of freedom is negligible.

(ii) The molecules are spherically symmetrical either actually or by virtue of free-rotation.

(iii) The inter-molecular degrees of freedom are independent of the density.

(iv) The potential energy of a pair of molecules is of the form given by de Boer.

The (i) condition excludes all the isotopes of the lightest molecules hydrogen and helium, whereas the polar molecules are ruled out by conditions (ii) to (iv).

However it is not possible to divide substances into those for which principle holds and those for which it does not hold good i.e. Different substances accord with the principle to a greater or lesser degree. Heavy inert elements follow the principle to a great extent whereas highly polar molecules, molecules forming hydrogen bonds and chemically reactive molecules show a large deviation from the principle of corresponding states.

Suggestion has applied the principle of corresponding states to examine.

(i) The critical parameters

(ii) The Virial coefficients and the Boyles temperature

(iii) The densities of coexistent phases

(iv) The vapour pressure, heat capacity, coefficient of thermal expansion and surface tension of the liquid phase

(v) The temperature and the pressure at the triple point.

The following gases follow the principle:

Argon, Krypton, Xenon, Neon (less accuracy).

The following gases follow the principle with fair accuracy:

Nitrogen, oxygen, carbon dioxide, methane.

de Boer discussed this law from the point of view of quantum theory. He introduced new parameters, characteristic energy (ε) and characteristics length (σ) for each molecular species.

The macroscopic quantities temperature, pressure and volume in terms of molecular units are $\frac{\varepsilon}{\kappa}, \frac{\varepsilon}{\delta^3}$ and N σ^3 respectively where K is

Boltzmann's constant and N is Avogadro's number of particles.

According to law of corresponding states then there should be a unique relation between pressure, volume and temperature expressed in molecular units. It has been found that the relation which is obtained depends also on a parameter.

$$\Lambda^* = h/\sigma \cdot \sqrt{m\varepsilon}$$

which has a characteristics value for each substance considered.

The classical law of corresponding states is obtained by the limiting value $A^* \rightarrow 0$.

Various properties of gaseous elements have been examined as this basis and it has been formed that, for higher elements, the molar volume and internal energy depart from the classical limit. This derivation can be related to zero-point energy.

The solid-liquid transition has also been considered and an explanation has been provided of the fact that a helium isotope of mass 4 cannot be solidified under its equilibrium vapour pressure.

de Boer made calculations of the molar volume, as a function of temperature, of the helium isotope of mass 3 in the liquid and the gaseous states and the equilibrium vapour pressure of the liquid. The vapour pressure curve agrees fairly accurately with the predictions made by de Boers. More recently various other measurements such as those of density function, viscosity etc. have been made for this isotope of helium and the values are found to agree with the predicted value at 4.2 K. The values are about 12% lower at 1.3 K.

The law of corresponding states has now been extended to solids at low temperatures, considering the frozen rare gases. This method has been used to calculate the potential energy and the interatomic distances of Ne, Ar, Kr and Xe. The values are found to be in agreement with experimental values..

4.7 The Critical Phenomenon and Critical Constants

All gases are composed of molecules and can be liquified by increase of pressure and/or decrease of temperature. However in the liquifaction of gases temperature plays more important part than pressure. It is possible to liquify all gases at atmospheric pressure but some of them can not be liquified at room temperature.

Thus for every gas there is a temperature above which it can not be liquified by increase of pressure. This is known as critical temperature.

Various critical constants have been defined as:

(i) **Critical temperature of a gas (T_e).** It is the minimum temperature which must be reduced before it can be liquified by the application of pressure. A gas cannot be liquified by pressure above its critical temperature, no matter how high the applied pressure may be. Each gas possesses a definite critical temperature, *e.g.,* critical temperature of CO_2 is 31.1°C.

(ii) **Critical pressure of a gas (P_e).** It is the minimum pressure necessary for the liquifaction of a gas at critical temperature.

(iii) **Critical volume of a gas (V_e).** It is the volume occupied by 1 gm. mole of gas or liquid under critical pressure and under critical temperature.

Critical constants of some gases are given in Table 4.1.

Table 4.1
Critical Constants of Gases

Gas	Tc (°A)	Pc (Atm)	Vc (cc)
Helium	5.20	2.26	61.55
Hydrogen	33.20	12.80	69.70
Oxygen	154.28	49.70	74.40
Nitrogen	125.97	33.50	90.00
CO_2	304.15	73.00	94.20
Cl_2	417.00	76.10	123.50
NH_3	405.50	111.50	72.00

4.8 Isotherms of Carbon Dioxide

The curves representing the variation of pressure and volume at constant temperature are known as *isotherms.* In the case of an ideal gas the isotherms would be rectangular hyperbolas.

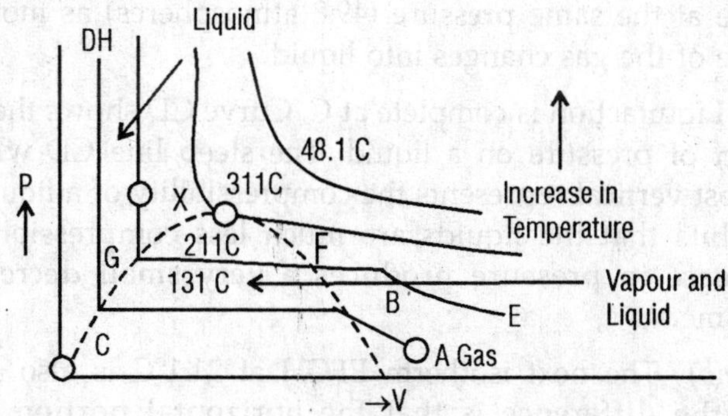

Fig. 4.2.

The effect of pressure on volume of carbon dioxide was studied by *T*. Andrews in 1896. He enclosed carbon dioxide gas in a capillary tube and measured the volume occupied by the gas at different pressures upto about 400 atmospheres.

On plotting volume reading against the pressure, a set of the curves of the type shown in Fig. 4.2 were obtained. Each curve corresponds to one particular temperature and is, therefore, called isotherm. The curves should be rectangular hyperbolas like that expected for an ideal gas. It is quite clear that CO_2 does give rectangular hyperbolas although with increasing temperature the curves approximate to one. We consider the various isotherms separately.

(*a*) **At 13.1°C.** The isotherm reveals that carbon dioxide is entirely gaseous at low pressure (*e.g.*, along AB) and the volume decreases with increasing pressure in the normal fashion (*i.e.*, AB is roughly hyperboloid). The point A corresponds to gaseous CO_2 occupying a certain volume under a certain pressure. At low pressures, when the pressure is increased, the volume decreases as depicted by the curve AB.

At B liquifaction starts; vapour is converted into liquid of very much greater density and the volume decreases suddenly since the volume occupied by a liquid has been much less than that of gas so the BC is almost parallel to the volume axis indicating the sudden decrease in volume taking place at the same pressure (49.8 atmospheres) as more and more of the gas changes into liquid.

Liquifaction is complete at C. Curve CD shows the small effect of pressure on a liquid. The steep line CD which is almost vertical represents the compressibility of a liquid and exhibits that the liquids are much less compressibly, *i.e.*, increase in pressure produces a very small decrease in volume.

(*b*) The next isotherm EFGH at 21.1°C is also similar but the difference is that the horizontal portion where liquifaction takes place is slightly short (*i.e.*, the range of volumes over which gas and liquid can co-exist is smaller) and the vapour pressure higher (*i.e.*, the pressure at which liquifaction commences is higher).

(*c*) As the temperature is raised, this horizontal portion becomes less and less till at 31.1°C it reduces almost to a point X. Above this temperature there is no trace of horizontal portion, *i.e.*, beyond 31.1°C no indication of liquifaction is there at all.

Andrews applied a pressure of 300 to 400 atmospheres but could not liquify CO_2 if the temperature was beyond 31.1°C. Below 31.1°C a pressure of 75 atmospheres was quite sufficient to liquify carbon dioxide.

Point X is the critical point for CO_2. The distinction between liquid and vapour state disappears at this point and CO_2 exists in a state called the **Critical state**. The isotherm passing through critical point is called the **critical isotherm.**

4.9 Continuity of State

Although the transition from liquid to vapour and *vice versa* generally takes place suddenly and the two states can be sharply distinguished from each other as is seen from AB and CD parts of 13.1°C isotherm for carbon dioxide (Fig.

4.2), yet there is a possibility of passing from one state of aggregation to the other without sudden change taking place. Therefore, for further theoretical considerations it is important to show that the liquid state is not a sharp and discontinuous transition from gaseous state but rather a continuation of the gaseous phase into the region of very strong intermolecular attractions where the volume is very small. The idea of a continuity from liquid to gaseous state will be understood from the following consideration:

The dome-shaped curve represented by the dashed line in Fig. 4.2, connects the end points of the horizontal portions of the isotherms. But a horizontal portion represents two phases in equilibrium. Therefore, all points inside the dome correspond to equilibrium between liquid and vapour. For such points it is always possible to draw any sharp distinction between liquid and vapour because there is a definite boundary surface between the two phases.

For points lying outside the dome only one phase (liquid or vapour) is present, and here it is not possible to draw any sharp distinction between liquid and vapour. This fact is called the *principle of continuity of states* and may be demonstrated with reference to Fig. 4.2.

4.10 Vander Waal Constants and Critical Constants

Ans. Vander Waals equation for one mole of gas is

$$\left(P+\frac{a}{V^2}\right)(V-b) = RT$$

on expending and rearranging the equation on descending power of V, we get

$$PV^3 - V^2 (RT + Pb)+ aV - ab = 0$$

Division of the whole equation by P gives

$$V^3-V^2 \left(b+\frac{RT}{P}\right)+V\left(\frac{a}{P}\right)-\frac{ab}{P}=0 .$$

This is a cubical equation in V which when solved for given values of P and T will yield either three values of V or one real and the other two imaginary.

At the critical point, the three value of V are not only real and positive but also identical and equal to the critical volume. Mathematically we have,

$$(V-V_c)^3 = 0 \quad \text{...(1)}$$

Expending this equation by the binomial theorem we get

$$V^3-(3V_c)V^2-(3V_c^2)V-V_c^3 = 0 \quad \text{...(2)}$$

Substituting P_c and T_c for P and T respectively in equation (1) we get,

$$V^3-V^2\left(b+\frac{RT_c}{P_c}\right)+V\left(\frac{a}{P_c}\right)-\frac{ab}{P_c}=0 \quad \text{...(3)}$$

The equation (3) should be identical with equation (2) and the coefficients of like term must be equal. Equating the coefficients of like powers of V in equations (2) and (3), we get

$$3V_c = b+\frac{RT_c}{P_c} \quad \text{...(4)}$$

$$3V_c^2 = \frac{a}{P_c} \quad \text{...(5)}$$

and

$$V_c^3 = \frac{ab}{P_c} \quad \text{...(6)}$$

Dividing the equation (6) by equation (5) we get

$$V_c = 3b \quad \text{...(7)}$$

Substituting the value of V_c in equation (5), we get

$$3(3b)^2 = \frac{a}{P_c}, \quad \text{or} \quad P_c=\frac{a}{27b^2} \quad \text{...(8)}$$

Substituting the values of V_c and P_c from equations (7) and (8) respectively, in equation (5),

$$3\times 3b = b+\frac{RT_c}{a}\times 27b^2$$

or

$$T_c = \frac{8a}{27Rb} \quad \text{...(9)}$$

From equation (1),

$$a = P_c \times 27b^2$$

Substituting this value in the equation (9),

$$T_c = \frac{8 \times P_c \times 27b^2}{27Rb}$$

or $$b = \frac{RT_c}{8P_c} \quad \text{...(10)}$$

Substituting this value of b in equation (9) yields

$$T_c = \frac{8a \times 8p_c}{27R \times RT_c}$$

$$a = \frac{27R^2T_{c^2}}{64P_c} \quad \text{...(11)}$$

Combination of Equations (5) and (7) leads to the value of R in terms of critical constants

$$R = \frac{8P_cV_c}{3T_c} = 2.67\frac{P_cV_c}{T_c} \quad \text{...(12)}$$

NUMERICALS

Q. *Example one mole of ammonia was found to occupy 25 dm^3 at 300 K. Calculate the pressure that would have been exerted by it using.*

(i) the ideal gas equation.

(ii) the Vander Waals equation.

[For NH_3, $a = 4.0$ atm dm^6 mol^{-2} and $b = 0.036$ dm^3 mol^{-1}]

Solution. (*i*) For an ideal gas

$$PV = RT$$

or $$P = \frac{RT}{V}.$$

Here

$$R = 0.082 \text{ atm dm}^2 \text{ K}^{-1} \text{ mol}^{-1}$$
$$T = 300\text{K}$$

and $$V = 2.5 \text{ dm}^3$$

$$\therefore \quad P = \frac{0.082 \times 300}{2.5}$$
$$= 9.84 \text{ atm. } \textbf{Ans.}$$

(*ii*) The Vander Waals equation for 1 mole is

$$\left(P+\frac{a}{V}\right)(V-b) = RT$$

or $$P+\frac{a}{V^2} = \frac{RT}{V-b}$$

or $$P = \frac{RT}{V-b}-\frac{a}{V^2}.$$

Here $R = 0.082$ atm dm^3 K^{-1} mol^{-1}

$T = 300$ K

$V = 2.5$ dm^3

$a = 4.0$ atm dm^6 mol^{-2}

$b = 0.036$ dm^3 mol^{-1}

$$P = \frac{0.082\times 300}{2.5-0.036}-\frac{4.0}{(2.5)^2}$$

$$= \frac{24.6}{2.464}-0.640$$

$= 9.984 - 0.640 = 9.344$ atm. **Ans.**

4.11 The Law of Corresponding States and the Reduced Equation of States.

Vander Waals in 1881 said that if the pressure, volume and temperature of a gas are expressed as fractions of critical constants in the equation, an important law called the law of the corresponding states is obtained.

Suppose: $\pi = \frac{P}{P_c}$ or $P = \pi P_c$

$\phi = \frac{V}{V_c}$ or $V = \phi V_c$

$\theta = \frac{T}{T_c}$ or $T = \theta T_c$

π, ϕ and θ are called the **reduced pressure**, the **reduced volume** and the **reduced temperature** respectively.

Substitution these values of *P, V* and *T* in the Vander Waals equation

$$\left(P+\frac{a}{V^2}\right)(V-b) = RT, \text{ we get}$$

$$\left(\pi P_c+\frac{a}{\phi_2 V_c^{\ 2}}\right)(\phi V_c - b) = R\theta T_c$$

Since $P_c = \dfrac{a}{27b^2}, V_c = 3b$ and $T_c = \dfrac{8a}{27Rb}$, the above equation takes the form

$$\left(\pi\frac{a}{27b^2}+\frac{a}{9\phi_2 b^2}\right)(3\phi b - b) = R\theta\frac{8a}{27Rb}$$

or
$$\frac{a}{27b_2}\left(\pi+\frac{3}{\phi_2}\right)b(3\phi-1) = \frac{a}{27b}.8\theta$$

or
$$\left(\pi+\frac{3}{\phi_2}\right)(3\phi-1) = 8\theta .$$

This equation is known as the **Vander Waals** reduced **equation of state**. This equation does not involve either *R* or the Vander Waals constants *a* and *b* which are characteristic of the given gas. Hence it is a general equation and applicable to all liquids and gases.

It is evident from the equation that when two or more substances have the same reduced temperature and reduced pressure, they will have the same reduced volume. Thus two or more substances under these conditions are said to be in *corresponding states* and the above statement is referred to as *the law of corresponding states.*

Q. 1. *The critical pressure and temperature of carbon dioxide are 73 atm. and 31°C respectively. Calculate the values of Vander Waals constants. [R = 0.082 dm^3 atm K^- mol^{-1}].*

Solution. It is known that

$$R = \frac{8}{3}\times\frac{P_c V_c}{T_c}$$

or $$V_c = \frac{3}{8} \times R \frac{T_c}{P_c}$$

Now $$R = 0.082 \text{ dm}^3 \text{ atm K}^{-1} \text{ mol}^{-1}$$
$$T_c = 273 + 31 = 304 \text{ K}$$
$$P_c = 73 \text{ atm.}$$
$$V_c = \frac{3}{8} \times \frac{0.082}{1} \div \frac{304}{73}$$
$$= 0.128 \text{ dm}^3 \text{ mol}^{-1}$$

Now $$V_c = 3b$$

or $$b = \frac{V_c}{3} = \frac{0.121}{3}$$
$$= 0.0427 \text{ dm}^3 \text{ mol}^{-1}. \textbf{ Ans.}$$

and $$a = 3V_c^2 P_c$$
$$= 3 \times (0.128)^2 \times 73$$
$$= 3.59 \text{ dm}^6 \text{ atm mol}^{-2}. \textbf{ Ans.}$$

Q. 2. *Calculate the Vander Waals constants, if $T_c = 33.3$ K, $P_c = 128$ atm and $V_c = 0.065$ dm³ mol⁻¹, for hydrogen. [R = 0.08205 dm³ atm K⁻¹ mol⁻¹].*

Solution. It is know that

$$a = 3V_c^2 P_c$$
$$b = \frac{V_c}{3}.$$

Here $$P_c = 12.8 \text{ atm}$$
$$V_c = 0.065 \text{ dm}^2 \text{ mol}^{-1}$$
$$a = 3 \times (0.065)^2 \times 12.8$$
$$= 0.1622 \text{ dm}^6 \text{ atm mol}^{-2}. \textbf{ Ans.}$$

and $$b = \frac{0.065}{3}$$
$$= 0.02167 \text{ dm}^2 \text{ mol}^{-1}. \textbf{ Ans.}$$

Q. 3. *Vander Waals constants for carbon dioxide are $a = 3.6$ litre² atm mol⁻¹ and $b = 0.0428$ litre atm mol⁻¹. Calculate T_c. [R = 0.982 litre atm deg⁻¹ mol⁻¹]*

Solution. We know that

$$T_c = \frac{8a}{27Rb}$$

Now $a = 3.6$ litre^2 atm mol^{-2}

$b = 0.0428$ litre atm mol^{-1}

$R = 0.082$ litre atm deg^{-1} mole^{-1}

$$\therefore \quad T_c = \frac{8 \times 3.6}{27 \times 0.082 \times 0.0428}$$

$$= 304.5 \text{ K}$$

Q. 4. *Calculate the reduced pressure, volume and temperature of methane gas when 1 mole of gas is contained in a 5 litre flask under 5 atm. pressure. What will be temperature of the gas when*

$T_c = 190.2$
$V_c = 0.0988$ *litre mol*$^{-1}$
$P_c = 45.6?$

Solution. We know

Reduced pressure $\pi = P/P_c$

Here $P = 5$ atm. and $P_c = 45.6$ atm.

Substituting the values,

$$\pi = 5/45.6 = 0.1096$$

Reduced volume $\phi = V/V_c$

Here $V = 5$ litres and $V_c = 0.0988$ litre/mole

Substituting the values, we get

$$\phi = \frac{5}{0.0988} = 50.607$$

According to the law of corresponding states,

$$\left(\pi + \frac{3}{\phi^2}\right) \times (3\phi - 1) = 8\theta$$

where θ is the reduced temperature

i.e.,
$$\theta = \frac{\left(\pi + \frac{3}{\phi^2}\right)(3\phi - 1)}{8}$$

Substituting the values of π and ϕ.

$$= \frac{\left[0.1096 + \frac{3}{(50.607)^2}\right](3 \times 50.607 - 1)}{8}$$

$$= 2.0888$$

We know that reduced temperature,

$$\theta = \frac{T}{T_c}$$

$$T = T \times \theta$$

Her $\quad T_c = 190.2\text{ K}$

Substituting the values.

Temperature $= 190.2 \times 2.0888 = 397.3\ K$. **Ans.**

4.12 LIQUIFACTION OF HELIUM

In 1908, the last of the permanent gases, helium, was liquified by Onnes. Making use of the equation of state for helium, he concluded that it was possible to liquify helium by cooling it below its inversion temperature (50.5 K) in hydrogen boiling under reduced pressure.

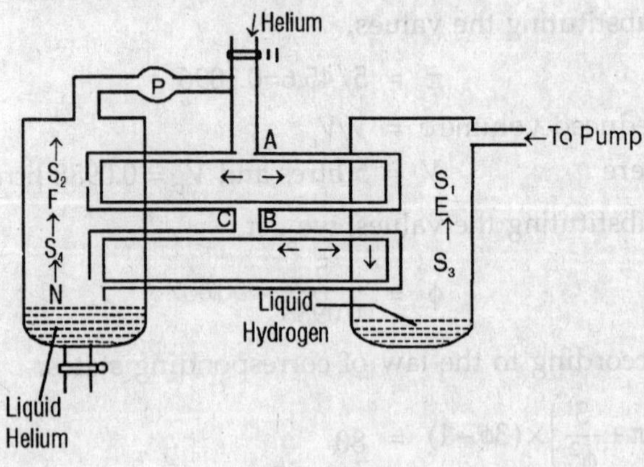

Fig. 4.3

The apparatus used by onnes for liquifaction of helium was similar to one used for liquifaction of hydrogen but smaller than the one used for hydrogen. The apparatus is shown in Fig. 4.3.

Helium is compressed to a pressure of about 40 atmospheres and is passed through the spiral tube at A where it gets divided into two portions. One portion of Helium passes through the spiral S_1 in E and it gets cooled as it is surrounded by hydrogen boiling under reduced pressure. The other portion of the Helium gas that passes through coil S_2 in F gets cooled due to outgoing cooled Helium gas. Similar processes spirals in the S_3 and S_4. The process gets repeated again and again and when the temperature of Helium is quite low, it gets liquified after passing through the nozzle N where cooling occurs due to Joule-Thomson effect. This outgoing Helium is compressed again by pump P and fed back of spirals S_1, S_2, S_3, and S_4. Liquified helium is collected in Dewar flask.

4.13 Helium I and II

Helium was the last gas to be liquified on account of having the lowest critical temperature (–268°C) of all known gases. It is a colourless transparent very volatile liquid and has the lowest boiling point of 4.2K at a pressure of 1 atmosphere. It has no normal freezing point and shows quite a few other properties which single it out as almost an exceptional and unique substance.

Solid helium can not be obtained merely by lowering the temperature of the liquid. Kamerlingh Onnes failed to solidify helium inspite of the fact that he reached a temperature of 0.84 K. Solid helium was first obtained by Keesom by subjecting liquid helium to very high pressures. The solid can be obtained at a pressure of 250 atmospheres at 4.2K, while only under 23 atmospheres at 1.1K. Thus it is necessary to increase the pressure simultaneously while lowering the temperature of the liquid. Further investigations showed that at a high pressure, solid helium could be obtained in equilibrium with the vapour at temperature well above the critical point of the gas. Thus at 5800 atmospheres, solid helium is obtainable at a temperature as high as 42K. This is the curious property of the helium system that although liquid can not exist above the critical temperature (5.2K) but solid can exist if sufficiently high pressure is applied. Hence at high pressures, the melting point exceeds the critical temperature and so the solid helium 'melts' to form the gas.

The phase diagram of helium system is as shown in Fig. 4.4. in which all portions are not on the same scale. A study of the phase diagram shows that it is entirely different from that of all other substances. The fusion curve (or the solid liquid phase line) and the saturated vapour pressure curve (or the liquid vapour phase line) do not meet in a point, as in the case of other systems and if the vapour pressure curve is pursued down to lower temperatures it is found that the vapour and the liquid continue in equilibrium even up to the absolute zero. Thus the three phases solid, liquid and vapour are never found to coexist or in other words *helium has no triple point.* This certainly is an extraordinary thing and it shows that helium is a unique liquid.

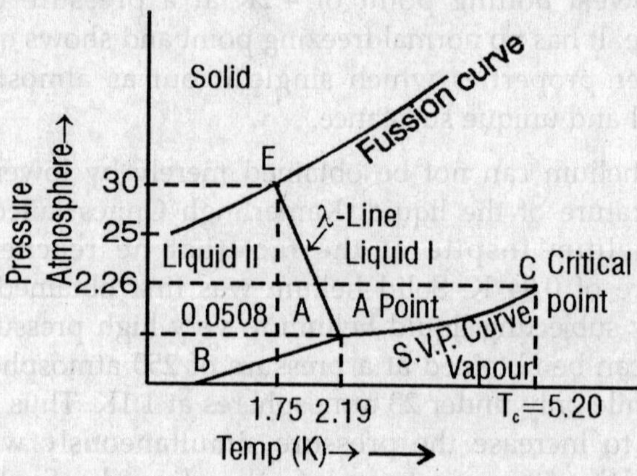

Fig. 4.4

The curve EFK (Fig. 4.4) shows the effect of pressure on the melting point. It proceeds upwards to the right, takes a sharp bend at *E* and flattens out at low temperatures meeting the pressure axis at 25 atmospheres, which indicates that *helium will not solidify even at 0K if it is not subjected to pressures exceeding 25 atmospheres*. The S.V.P. curve on the other hand, seems to proceed normally to the left towards the origin (p = 0, T = 0) but to the right it terminates at critical point C corresponding to a temperature of 5.2K and a pressure of 2.25 atmospheres. The point A (2.19K) is called the λ *point* of liquid helium under its own vapour pressure.

For helium in the liquid phase, there is a *phase transition* known as the λ-transition, which divides the liquid state into two phases, helium I and helium II. The fusion curve and S.V.P. curve are joined by the λ-line that runs between the point E (T = 175 K, p = 30 atoms) and A (T = 2.19K and p = 0.05 atoms) with helium I to its right and helium II to its left. Thus helium is present in the liquid form on either side of the 1-line. Kamerlingh Onnes, found that liquid helium shows an extremely interesting behaviour if it is cooled below its boiling point (4.2K) to about 2.18K. He observed that the density passes through an abrupt maximum at 2.19K decreasing slightly after that as shown in Fig. 4.5, the density first rises with the fall of temperature from 4.2K upto 2.19K reaching a maximum value of .1462 at 2.19K and then decreases with the decrease of temperature. Below 2.19K, the liquid helium which was contracting when cooled now begins to expand.

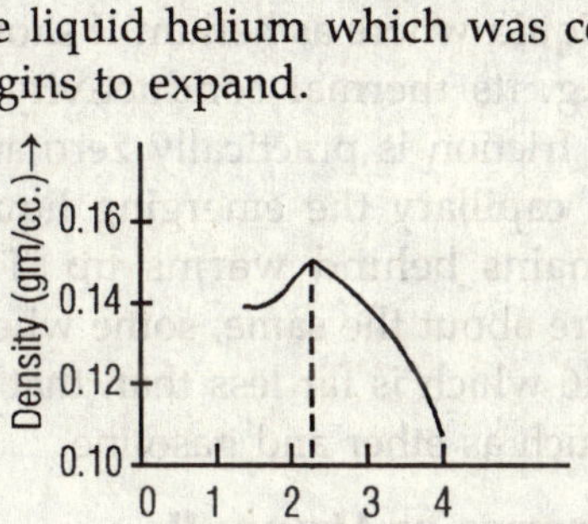

Fig. 4.5

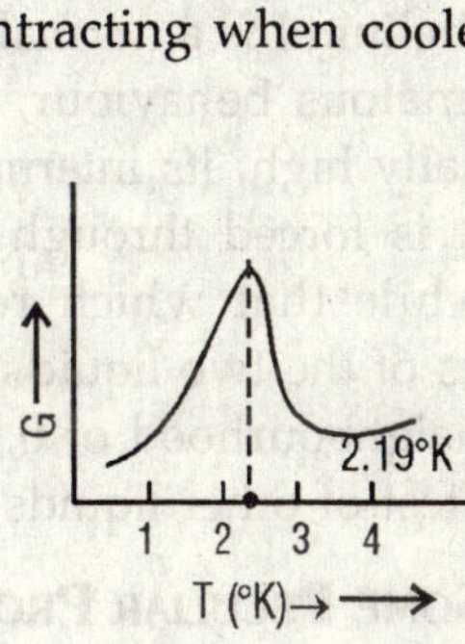

Fig. 4.6

The specific heat of liquid helium at constant volume has been plotted in Fig. 4.6. The specific heat increases, upto 2.19K and at this temperature there is a sudden and abnormal increases in its value. Beyond 2.19K the specific heat first decreases and then increases. The dielectric constant also behaves in a similar manner. The specific heat temperature graph at 2.19K appears like the Greek letter lamda (λ) and thus this temperature (2.19K) at which specific heat changes abruptly is known as 1–point. There is a basic difference between the natures of the liquid above the below the λ–point (2.19K). Liquid helium above 2.19K behaves in a normal way and is known as liquid helium I and helium existing below this temperature is the liquid helium II due to its abnormal properties. No heat is involved or absorbed during the transition from one form of helium to another. Therefore, no latent heat is involved in the transition He I $\rightarrow$ He II, it shows that (a) the entropy is continuous across the curve, i.e., the entropy of He II is practically the same as that of the He I, and (b) there is no change of density during transitions, i.e., the density of both type of liquids is about the same.

While the viscosity of liquids increases with decrease in temperature, that of liquid helium I decreases. In this respect He I resembles a gas. The viscosity of He II is almost zero and it can flow rapidly through narrow capillaries. Liquid helium I is a normal liquid where as helium II shows an anomalous behaviour, e.g. its thermal conductivity is abnormally high, its internal friction is practically zero and when it is forced through a capillary the emerging liquid cools while that which remains behind warms up. The densities of the two liquids are about the same, some where in the neighbourhood of 0.146 which is far less than that of the lightest of other liquids such as ether and gasoline.

4.14 Some Peculiar Properties of Helium II

In many of its properties, the behaviour of liquid helium II is quite different from other substances. It behaves strangely and shows some very unique properties. Some of these:—

(i) *Extreme fluidity*. A significant property of liquid helium II is its viscosity which is abnormally low even less than that of hydrogen gas and its rate of flow through capillaries is independent of the pressure difference across them. According to Poiseuile's law, the rate of a liquid through a capillary tube or a annular space is inversely proportional to the viscosity. If the rate of flow of liquid helium through a fine annulus is measured as a function of temperature, it is found that at λ-point the rate of flow is extraordinarily large.

The viscosity decreases by a factor of about ten when it passes through the l-point. Kapitza found that it is given by the following relation

$$\frac{\eta \text{ for liquid He II}}{\eta \text{ for liquid He I}} = 10^{-3}$$

Thus liquid He II has particularly zero viscosity and can flow rapidly, without any resistance, through narrow tubes. Because of this the liquid He II is known as a superfluid, a new frictionless state of matter. The properties associated with it have been characterized as *superfluidity*.

(ii) *High Thermal conductivity*. Thermal conductivity of liquid helium I is normal and is of the order of 10^{-4}, which is of the same order as for gases at ordinary temperatures. However He II has an extraordinarily high coefficient of thermal conductivity. The heat transported per unit temperature gradient is several hundred times as great as that in copper at room temperature. He II is about 800 times more conducting than copper and about 13.5×10^6 times more conductivity than liquid He I, the absolute value of its thermal conductivity is 820 cals. cm^{-1} deg^{-1} sec^{-1}. Its heat flow is also not proportional to the temperature gradient.

Daunt and Mendelssohn, in 1947, observed the striking analogy between the unimpeded flow of helium atoms in liquid helium II and the phenomenon of *super-conductivity*. No trace of electrical resistance is observed in a material which is in a super conducting state i.e., the electrons move freely in the conductor, unimpeded by any electrical resistance.

(iii) *Formation of films over solid-surfaces.* Liquid helium can creap along solid surfaces in the form of highly mobile film generally called the Rollin (also called Rollin-Simon) film. This film is of the order of several hundred atoms thick. The properties of the film were investigated by Daunt and Mendelssohn who observed the following remarkable behaviour of helium.

When a tube containing helium II is placed in a helium II bath, it is found that:

(a) In case the liquid level inside the tube is lower than that outside, the liquid helium from outside starts creeping into the tube along its surface in the direction of arrows shown in Fig. 4.7 (a) the process continues till the levels inside and outside the tube, become equal.

(b) In case the tube be raised and the level of helium inside it is higher than that outside it, the liquid from inside the tube starts creeping out of it along its surface in the direction of arrows till the two levels again become equal.

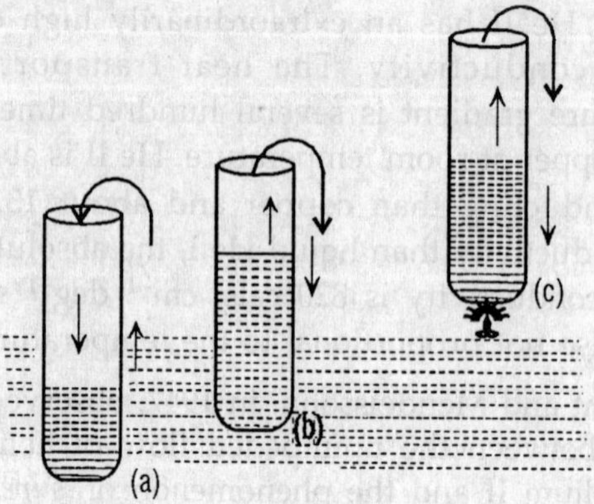

Fig. 4.6

(c) In case the tube is lifted entirely out of the He bath, the liquid inside the tube now creeps out along the surface of the tube, collects at its bottom in the form of drops and falls into the liquid below till the whole liquid inside the tube is drained out [Fig. 4.6] and the tube is empty.

All these experiments show that the liquid helium II attains the lowest level, but the rate of transfer of the liquid is independent of the difference in level and the nature of the surface. It, however, depends upon the temperature and the perimeter of the surface over which it has to pass.

Thus liquid He II appears to defy gravity by creeping out of the containing vessel by coating the walls with a thin film of the liquid.

(iv) *The "Fountain' and 'Mechano caloric' Effect.* Another peculiar property of liquid He II shown is in the "Fountain' and the 'Mechano-caloric' effect. The fountain effect, which is illustrated in Fig. 4.8. It shows that heat locally supplied to the system produces a pressure difference. The phenomenon was discovered by Allen and Jones in 1938 but the apparatus shown in Fig. 4.7 is one due to Allen and Misener. In the apparatus, He II is taken in a tube AB which is open at both the ends and kept inside a Dewar flask which also contains liquid helium II. The upper part A of the tube is in the form of a capillary and kept projecting out of the liquid bath the lower part B having a small opening 0 at the bottom is packed with fine emery powder. The interspaces between the powder particles provide fine capillaries for the liquid He II to flow into the tube. When heat radiations from an ordinary torch are allowed to fall on the powder, liquid He is found to spurt out of the capillary tube in the form of a fountain which has been observed as high as 30 cm. This is a consequence of absorption of energy by the energy powder.

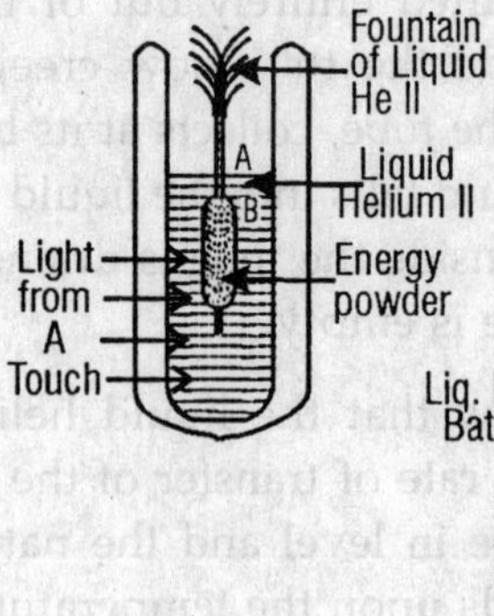

Fig. 4.8

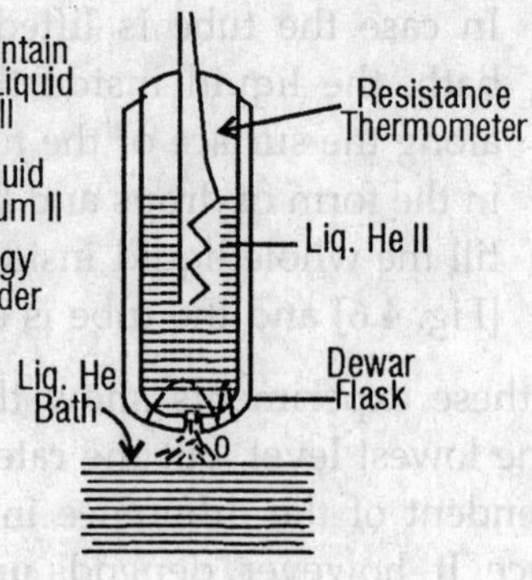

Fig. 4.9

The mechano caloric effect is shown in Fig. 4.9. This effect is opposite to the fountain effect and shows that the flow of liquid gives rise to a temperature difference. A Dewar flask is filled partly with the liquid helium II and provided with a small hole O at the bottom. The hole is plugged by fine powder. A resistance thermometer is introduced in the Dewar flask and then its top is closed. If now the flask is above the level of the liquid He bath, the liquid from the flask drains out through the powder and its temperature is found to increase by about 0.1°C. On the other hand if the Dewar flask is dipped into the bath and the liquid helium is allowed to flow into it through the opening O via the powder, the temperature inside the flask is observed to decrease.

To account for the peculiar properties of liquid helium II, it is generally considered as a mixture of two different kinds of liquids. One of these is made up of normal atoms with normal viscosity while the other is made up of the atoms which are in the state of lowest energy (zero-point energy) with zero entropy and are capable of moving through the normal atoms without any friction of viscosity. According to the present day views, this part of the liquid is of 'condensed' atoms and makes its first appearance at λ point. Such a liquid is known as *'superfluid'* and its property to flow through capillary tubes without showing any resistance is known as *superfluidity*. On lowering the temperature, the superfluid grows in quantity at the expense of normal component until at absolute zero the whole liquid consists of superfluid.

5

Liquifaction of Gases

5.1 Introduction

Liquifaction of Gases

Higher pressure bring the gas molecules together, the effect of attractive forces become appreciable. The attractive forces are opposed by motion of the gas molecules. At low temperature average K.E. of the molecules is low. Thus at extremes of these conditions gases liquify. Higher the temperature higher the pressure that is required to be applied to affect liquifaction. However for each gas there is a temperature above which it is impossible to liquify the gas no matter how high is the applied pressure. This temperature is called the *critical temperature of the gas.* The corresponding pressure required to be applied at critical temperature to liquify a gas is called *critical pressure.*

The Vander Waals equation is represented graphically as a family of isotherms (Fig. 5.1)

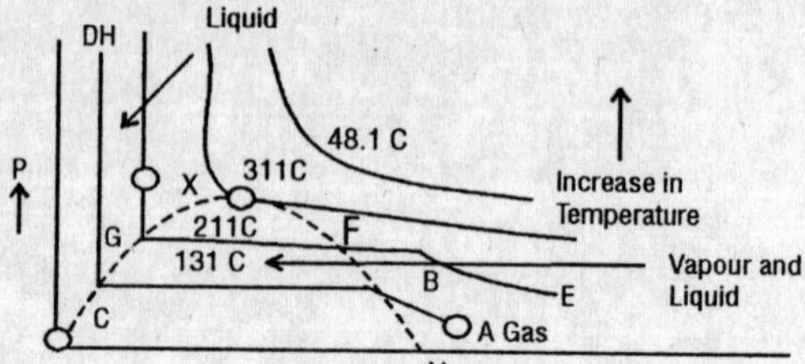

Fig. 5.1

Since the Vander Waals equation is cubic equation in V so V has three roots for any given value of P and T. At high temperature such as at T_2 only one root is real for all pressures. At lower temperatures such as T_1, all three roots are real over a certain pressure range. Thus a maximum and minimum are formed on the isotherm in the region where experimentally the phenomenon of condensation or vapourisation occurs. For some values of P and T the three real roots became identical and the isotherm is labelled as T_c which has a point of inflection at C with a horizontal tangent. This point is called the *critical point.*

Since the critical isotherm, has a point of inflection at the critical point, we have at this point.

$$\left(\frac{\partial P}{\partial V}\right)_{T_c} = 0 \quad \text{and} \quad \left(\frac{\partial^2 P}{\partial V^2}\right)_{T_c} = 0$$

Differentiating the Vander Waals equation and applying the results to critical point, we have

$$\left(\frac{\partial P}{\partial V}\right)_{T_c} = -\frac{RT_c}{(V_c - b)^2} + \frac{2a}{V_c^3} = 0$$

and

$$\left(\frac{\partial^2 P}{\partial V^2}\right)_{T_c} = \frac{2RT_c}{(V-b)^3} - \frac{6a}{V_c^4} = 0$$

Solving these two equations together with the Vander Waals equation as applied to the critical point, we obtain the following expression for the critical properties P_c, T_c, V_c in terms of Vander Waals constants a and b.

$$P_c = \frac{a}{27b^2} \quad \text{...(1)}$$

$$T_c = \frac{8a}{27Rb} \quad \text{...(2)}$$

$$V_c = 3b \quad \text{...(3)}$$

Conversely, the Vander Waals constants a and b can be expressed in terms of the critical properties. Thus the simultaneous solution of the above equations gives (1, 2, and 3)

$$a = \frac{27}{64}\frac{(RT_c)^2}{P_c} \qquad ...(4)$$

$$b = \frac{RT_c}{8P_c} \qquad ...(5)$$

Though Vander Waals equation describes the general behaviour of real gases, it holds only qualitatively and any attempt to test the equation quantitatively is doomed to failure one definite and clear deficiency is seen in the value of the ratio of $\frac{P_cV_c}{RT_c}$, which according to Vander Waals equation is $\frac{P_cV_c}{RT_c} = \frac{\left(\frac{a}{27b^2}\right)(3b)}{R\left(\frac{8a}{27Rb}\right)} = \frac{3}{8} = 0.375$ for all gases. But actual values of $\frac{P_cV_c}{RT_c}$ for different gases formed experimentally are different. In most cases they are much smaller than predicted value.

However this equation for real gas is not expected to hold good at high densities such as the critical region due to Vander Waals simplified reasoning. Testing of the equation for lower densities should give better correlation.

5.2 Principles of Liquifaction of Gases

According to kinetic theory of gases, all gases consist of tiny particles known as molecules. These molecules are moving with very high speeds in all possible directions. If the temperature of a gas in raised its K.E. increases and if the temperature is decreased the K.E. decreases. Due to decrease in K.E., the speed of molecules decreases and they move close to each other and thus the volume decreases. If we go on decreasing the temperature a stage will come when molecules come so close that they change into liquid state.

This change from gaseous to liquid state can be brought about by the increase in pressure. As already pointed out, all gases can not be liquified by the increase of pressure at room temperature. For every gas there is a certain temperature, called *critical temperature,* below which or to which the gas must be cooled and it is only then possible to liquify a gas by the application of pressure. This pressure required to liquify a gas at *critical temperature* (T_c) is called *critical pressure* (P_c)

It is thus concluded that the following condition should be satisfied for the liquifaction of gases:

(i) The temperature must be at or below the critical temperature.

(ii) A suitable pressure must be applied equal to critical pressure of the gas in at critical temperature. A smaller pressure will be sufficient if the gas is below its critical temperature.

It would be desirable if the gases are free from moisture, it would avoid formation of ice. In case of air it should be free from carbon dioxide as well.

5.3 METHODS OF LIQUIFACTION OF GASES

Any of the following methods may be used for the liquifaction of gases:

(i) By use of freezing mixtures

(ii) Rapid evaporation of volatile liquids

(iii) Joule Thomson effect

(iv) Adiabatic expansion of cold compressed gas

(v) Adiabatic demagnetisation.

Details of Methods

(1) *By the use of Freezing Mixtures:* This method was first to be used for liquifaction of gases. Faraday (1923) liquified SO_2, HCl, CO_2, H_2S etc. The apparatus used is shown in Fig. 5.2.

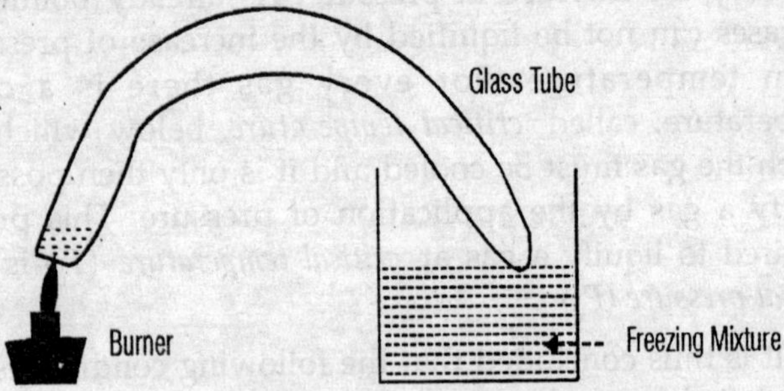

Fig. 5.2. Liquifaction by Faraday Method.

The apparatus consists of V-shaped hard glass tube. The gas was prepared in out of the arms by a suitable chemical reaction whereas the other arm was dipped in freezing mixture. The gas liquified under its own pressure, in this arm, because of external cooling. Some gases like H_2, N_2, O_2 etc. could not be liquified by Faraday and he called them *Permanent gases.*

Now we know that the case of so-called permanent gases the critical temperature is quite low and that is why they could not be liquified at room temperature.

Principle of Freezing Mixtures

By adding salt (sodium chloride) to water it is possible to attain a temperature much lower than 0°C. When salt is added to ice, the salt is dissolved in water adhering to ice and more ice melts. The heat required by ice to melt is extracted from the mixture and it results in a fall of the temperature of the mixture. This is the *principle of the freezing mixture.*

The lowest possible temperature that can be attained by making use of this principle of freezing mixtures depends upon the salt used in the process and is called the *eutectic temperature*.

The temperatures that can be attained by use of different salts are given in Table *5.1*.

Table 5.1

Salt	**Quantity of salt dissolved/100g of Mixture**	**Minimum Temp (°C)**
$MgSO_4$	19.0g	–3.9
NaCl	22.0g	–21.2
KCl	19.7g	–11.1
NH_4CL	18.6g	–15.8
$CaCl_2$	29.8g	–55.0
$NaNO_3$	37.0g	–18.5

The freezing mixture most commonly used these days is a mixture of solid carbon diomide (Dry ice) and ether. This mixture can produce temperature as low as –110°C.

5.4 BY RAPID EVAPORATION OF VOLATILE LIQUID

Evaporation causes cooling because during evaporation, liquid absorbs heat equal to its latent heat of evaporation, from its remaining part. Water was frojen by this process as far back as 1755. The 'cascade process' or pictet process for the liquifaction of gases.

In the Pictet process, which has been used by Dewar and by Onnes, the low temperature is reached step by step, that is why it is called 'cascade process'.

In this process a gas 'B' can be cooled below its critical temperature by rapidly evaporating another gas 'A' and thus the gas 'B' can be liquified. The rapid evaporation of 'B' was then used to liquify 'C' and so, e.g. SO_2 used to liquify CO_2 and CO_2 in turn was used to liquify O_2. It is not possible to liquify H_2 and He by this process because of very low value of their critical temperature and such a low temperature can not be attained by evaporating any other gas.

The critical temperature of oxygen is –118°C and the critical temperature of hydrogen is –240°C.

For liquifaction of hydrogen and helium Lindes' process must be used.

In the rapid evaporation process, the amount of cooling depends upon the *nature of the liquid* and the *rate of evaporation.*

These days the contrivance utilising the rapid evaporation process is used in a number of refrigerating machines. To produce very low temperatures generally liquified gases, e.g., ammonia, sulphur-dioxide etc. are used. Two types of such machines that are in used are (i) vapour compression machine, (ii) vapour absorption machine. The former type of machine are more efficient specially for large plants and need less initial cost, so they are more commonly used. The only difference between the two types machines lies in the manner of compressing the low pressure vapour.

Vapour compression machine. The schematic arrangement of vapour compression machine is shown in Fig. 5.3. It consists of the following three major parts:

(a) The compressor G.

(b) The condenser C.

(c) The evaporator E.

The compressor G is generally electrically driven and consists of a piston and two valves, one for the suction of low pressure vapour from the evaporator and the other valve is for the discharge of the compressed vapour to the condenser.

The condenser is made up of a coil of copper embedded in a vessel through which a coolant like cold water or brine is flowing.

When the piston of the compressor moves upwards, the pressure in the cylinder of compressor falls below the pressure in the evaporator, and due to this low pressure vapour is sucked through the valve and the suction pipe. When the piston moves downwards, the vapour is compressed and then delivered to the condenser C through the discharge valve and discharge pipe. The vapour at high pressure that enters the condenser maintained at low temperature liquifies in the condenser. The liquid passes through the expansion or regulator valve V which reduces the pressure of the liquid from the high pressure prevailing in the condenser to the low pressure in the evaporator. Because of the low pressure, the liquid evaporates taking its latent heat either from the cold storage space or the brine is cooled. The low pressure vapour is again sucked in by the compressor and the cycle of operations continues.

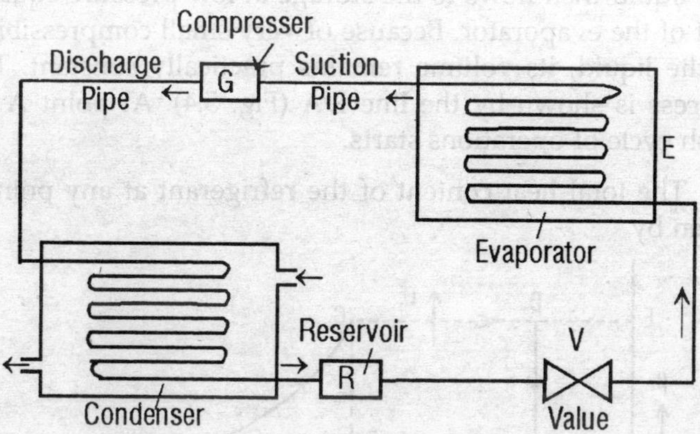

Fig. 5.3

The liquid used as a refrigerant should have the following characteristics :

(i) It should have a large latent heat of evaporation.

(ii) Its vapour should have small specific volume.

(iii) It should have sufficient vapour pressure at the temperatures of the evaporator.

Coefficient of Performance of a Vapour Compression Machine

The working of vapour compression machine can be represent on a *P–V*. diagram as shown in Fig. 5.4. The working liquid (refrigerant such as liquid NH_3, SO_2 etc. passes through the evaporation, it evaporates to vapours at constant pressure equal to that of saturated vapours absorbing the latent heat of evaporation from the objects to be cooled. This process can be represented by the line AB on *P–V* diagram (Fig. 5.4). Then the vapours are compressed adiabatically to a high pressure. This process is depicted by the curve BC (Fig. 5.4). At point C the pressure of the vapours is equal to that of the saturated vapour of the liquid at the temperature of the condenser. At this stage the heat of compression is removed at constant pressure by the coolant circulating the condenser so that the vapours are condensed to liquid phase. This process is represented by the line CD. The liquid then flows to the storage at low pressure equal to that of the evaporator. Because of very small compressibility of the liquid, its volume remains practically constant. This process is shown by the line DA (Fig. 5.4). At point A the fresh cycle of operations starts.

The total heat content of the refrigerant at any point is given by

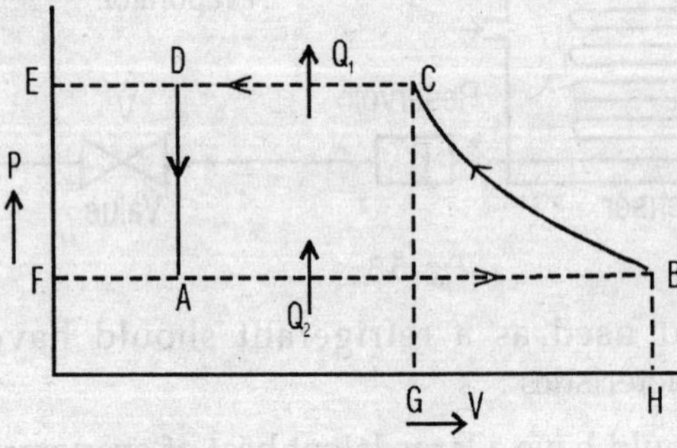

Fig. 5.4

$$H = U + PV, \quad \ldots(i)$$

where *U* is the internal energy,

so that

$$dH = dU + P\,dV + V\,dP \qquad \text{...}(ii)$$

According to First law of thermodynamics

Thus we can $\quad dQ = dU + PdV \qquad \text{...}(iii)$

$$dH = dQ + V\,dP \qquad \text{...}(iv)$$

Now the change from B to C is adiabatic, i.e., $dQ = 0$. Therefore, the area of the curve between BC and the pressure axis (i.e., ECBF) is given by

$$\int_B^C dH = \int_B^C V\,dP \qquad \text{(since } dQ = 0\text{)}$$

or

$$H_C - H_B = \int_B^C V\,dP$$

If V_1 is the constant volume of the liquid from D to A and P_2 and P_1 the pressures at points D and A respectively, the areas between the line AD and pressure axis, (i.e., EDAF) is $(P_2 - P_1)\,V_1$. Therefore, the work done in running the machine, is given by

$$\begin{aligned} W &= \text{Area ABCD} \\ &= \text{Area ECBF} - \text{Area EDAF} \\ &= \int_B^C V\,dP - (P_2 - P_1)\,V_1 \\ &= HC - HB - (P_2 - P_1)\,V_1 \\ &= HC - HB. \end{aligned}$$

Since for a given mass the volume of the liquid is negligible in comparison to the volume of its vapours. i.e.,

$$H_C - H_B \geq (P_2 - P_1)V_1 \qquad \text{...}(v)$$

The heat absorbed by the machine at low temperature is given by

$$Q_2 = \int dQ = \int (dU + PdV). \qquad \text{...}(vi)$$

But $\quad dH = dU + PdV + V\,dP.$

From A to B the change is isobaric, i.e., $dP = 0$, so that we have

$$dH = dU + P\,dV \qquad \text{...}(vii)$$

Comparing (vi) and (vii),

$$Q_2 = \int_A^B dH = H_B - H_A$$

Then the *Coefficient of Performance* of the vapour compression machine is given by

$$B = \frac{Q_2}{W} = \frac{H_B - H_A}{H_C - H_B} \quad \text{...(viii)}$$

5.5 By Joule-Thomson Effect

Introduction. The behaviour of real gases was thoroughly studied by Joule in collaboration with Thomson by performing a series of experiments (1852 to 1862) that led to the discovery of the phenomenon known as the "Joule Thomson effect." This may be stated as follows:

"When a gas under higher constant pressure is allowed to pass through a porous plug (or "throttle") to a region of lower pressure, its temperature is lowered."

Joule-Thomson Experiment. Joule and Thomson employed a long tube being thermally insulated from outside. They fitted it with a porous plug to allow the gas to be kept at different pressures P_1 and P_2 on the two sides of the tube. When the flow became steady, its temperature at the two extreme ends of the tube was recorded by using very sensitive thermometers and pressure difference was measured with a gauged.

Experimental Results. Various gases were subjected to the Joule-Thomson experiment and following results were reported:

(i) All the gases show a change in temperature on passing through the porous plug.

(ii) The change in temperature is found to be proportional to the difference in pressure on the two sides of the porous plug.

(iii) At sufficiently low temperature, all gases show a drop in temperature in passing through the porous plug. But at ordinary temperature, most gases suffer a fall in temperature while gases like hydrogen and helium record a rise in temperature.

(iv) Every gas has inversion temperature at a given initial pressure. Above this temperature, the gas shows heating effect. Below this temperature, the gas shows a fall in temperature.

Experimental set up. The experimental set up of Joule and Thomson is illustrated schematically. A tube thoroughly insulated (so as to be nearly under adiabatic conditions) was fitted with a porous plug, as indicated. The pressure on the left and right side of a porous plug were kept at P_1 and P_2 respectively ($P_1 > P_2$). By applying pressure on the piston *A* on the left slowly enough (so as not to change the pressure P_1), a volume of gas V_1 was forced slowly through the porous plug and then allowed to expand to the pressure P_2 and volume V_2 by moving the piston *B* on the right outward. While the expansion was taking place, accurate temperature readings were taken on the gas in the two chambers. A fall in temperature was observed in all cases except for hydrogen and helium.

Explanation*. The work done *on* the system at the left piston is P_1V_1 and the work done by the system at the right piston is P_2V_2. Hence the net work done by the system is:

$$w = P_2V_2 - P_1V_1$$

Since the process was conducted adiabatically.

$$q = 0$$

$$\therefore \quad \Delta E = -w$$

* A simple explanation of Joule-Thomson effect based upon the fact that the molecules of the gas are moving and so possess a definite amount of kinetic energy. When the gas expands, it does some work in overcoming the Vander Walls forces of attraction existing between the molecules. Thus a part of the energy is used up on doing this work and, therefore, the energy of the system falls and as result the temperature also falls.

or $$E_2 - E_1 = -(P_2V_2 - P_1V_1)$$

or $$E_2 + P_2V_2 = E_1 + P_1V_1$$

i.e., $$H_2 = H_1$$

$$dH = 0$$

Thus *during the free expansion of real gas, the heat content or enthalpy remains constant*. Under these conditions, the extent of the temperature change is found to depend both on the initial temperature and the pressure of the gas. Joule and Thomson defined a term, μ, known as *Joule-Thomson coefficient* as:

$$\mu = \left[\frac{\partial T}{\partial P}\right]_H$$

Thus μ *may be defined as the temperature change in degrees experienced by a gas a free expansion by a pressure drop of one atmosphere under conditions of constant enthalpy*. For cooling, μ is positive, while for an observed heating, μ is negative. The pressure at which the gas neither heats nor cools on expansion, i.e., where $\mu = 0$, is called an *Inversion Point*. This point varies with the temperature. It is observed that at a higher pressure, there is a temperature range within which μ is positive, i.e., cooling takes place. Below or above this range μ is negative, i.e., heating takes place (see Table 5.2 for values of μ for N_2).

Thus at the constant pressure, there are two inversion temperatures, an upper one and a lower one. Since most of the gases at room temperature are below their upper inversion temperature, hence they undergo cooling on adiabatic expansion. Hydrogen and helium on the other hand, are still above their upper inversion temperatures and hence they show heating. It may be expected that these gases would also exhibit cooling if their temperatures were lowered below their upper inversion point at a given pressure. This is actually the case.

Table 5.2
Joule Thomson Co-efficient For N_2

P.	μ at					
atms.	*-150°C*	*-100°C*	*0°C*	*100°C*	*200°C*	*300°C*
1	1.266	0.6490	0.2656	0.1292	0.0558	0.0140
20	1.125	0.5998	0.2694	0.1173	0.0472	0.0096
33.5	0.1704	0.5494	0.2377	0.1100	0.0430	0.0050
60	0.0601	0.4506	0.2088	0.0975	0.0372	0.0013
100	0.0202	0.2764	0.1679	0.0768	0.0262	0.0079
140	-0.0056	0.1373	0.1316	0.0582	0.0168	1.1122
180	-0.0211	0.0765	0.1015	0.0462	0.0094	1.0160
203	-0.0084	0.0084	0.0891	0.0419	0.0070	0.0171

Joule-Kelvin Effect for a Vander Waal's Gas

The Vander Waal's equation is

$$\left[P+\frac{a}{V_2}\right][V-b] = \text{RT}$$

In Vander Waal's theory it is assumed that the internal pressure arising from intermolecular forces between the molecules of a real gas is given by $\frac{a}{V^2}$ i.e.,

$$\text{Internal pressure} = \frac{a}{V^2}$$

Where a is constant. Therefore, the internal work done against these forces from V_1 to V_2 will be given by

$$w_2\int_{V_1}^{V_2}\frac{a}{V^2}\,dV = a\left[\frac{1}{V_1}-\frac{1}{V_2}\right] \qquad ...(3)$$

We have already proved that the external work done by the gas in the Joule-Thomson experiment is P_2V_2–P_1V_1. i.e.,

$$w_1 = P_2V_2 - P_2V_1$$

so the total work done by the gas is given by

$$w = w_1 + w_2$$

or

$$w = [P_2V_2 - P_1V_1] + a\left[\frac{1}{V_1} - \frac{1}{V_2}\right] \quad ...(4)$$

Since the system is thermally insulated, the external work must be done at the expense of its internal energy which is independent of temperature only i.e., $dE = w$.

Therefore, the decrease in internal energy (dE) is given by

$$dE = CV\,(T_1 - T_2)$$

where T_1 and T_2 are the temperatures on the high and low pressure sides respectively and C_v be the molar heat capacity of the gas at constant volume. When a state is approached, it therefore follows that

$$dE = w$$

$$C_v(T_1 - T_2) = (P_2V_2 - P_2V_1) + a\left[\frac{1}{V_1} - \frac{1}{V_2}\right]$$

$$C_v(T_1 - T_2) = (P_2V_2 - P_1V_2) + a\left[\frac{P_1}{RT_1} - \frac{P_2}{RT_2}\right] \quad ...(5)$$

$$(\because P_1V_1 = RT_1 \text{ and } P_2V_2 = RT_2)$$

On multiplication and neglecting a very small fraction $\frac{ab}{V^2}$, Vander Waal's equation takes the form

$$PV = RT - \frac{a}{V} + Pb$$

$$= RT - \frac{aP}{RT} + Pb \qquad [PV = RT]$$

Thus,

$$P_1V_1 = RT_1 - \frac{aP_1}{RT_1} + P_1b$$

and

$$P_2V_2 = RT_2 - \frac{aP_1}{RT_2} + P_2b$$

$$\text{or}\quad P_2V_2 - P_1V_1 = R(T_2 - T_1) + \left(\frac{aP_1}{RT_1} - \frac{aP_2}{RT_2}\right) + (P_2 - P_1)b$$

$$= R(T_2 - T_1) + a\left(\frac{P_1}{RT_1} - \frac{P_2}{RT_2}\right) + (P_2 - P_1)b$$

By putting this value in equation (5), we get

$$C_v(T_1 - T_2) = R(T_2 - T_1) + 2a\left(\frac{P_1}{RT_1} - \frac{P_2}{RT_2}\right) + b(P_2 - P_1) \quad \ldots(6)$$

As the fall in temperature is very small, we can put the term $\frac{2a}{RT_1}$ equal to $\frac{2a}{RT_2}$. Therefore, equation (6) takes the form

$$C_p(T_1 - T_2) = R(T_2 - T_1) + \frac{2a}{RT_1}(P_1 - P_2) + b(P_2 - P_1)$$

$$= R(T_2 - T_1) + (P_1 - P_2)\left(\frac{2a}{RT_1} - b\right)$$

$$\text{or}\quad C_p(T_1 - T_2) = (P_1 - P_2)\left(\frac{2a}{RT_1} - b\right) \quad \ldots(7)$$

$$[\because C_p - C_v = R]$$

$$\text{or}\quad \frac{T_1 - T_2}{P_1 - P_2} = \frac{1}{C_p}\left(\frac{2a}{RT_1} - b\right)$$

$$\text{or}\quad \frac{\Delta T}{\Delta P} = \frac{1}{C_p}\left(\frac{2a}{RT_1} - b\right)$$

$$\text{or}\quad \mu = \frac{1}{C_p}\left(\frac{2a}{RT_1} - b\right) \quad \ldots(8)$$

Where μ is the Joule-Thomson effect per unit pressure difference. From equation (8), it is evident

(i) at low temperature, $\frac{2a}{RT_1} > b; \mu > 0$ the gas will be cooled in Joule-Thomson expansion.

(ii) at high temperature, $\frac{2a}{RT_2} > b; \mu > 0$, the heating of the gas will take place.

(iii) at moderate temperature, $\frac{2a}{RT_2} > b; \mu > 0$ i.e., there is no change in temperature. This temperature is known as *temperature of inversion* and is usually devoted by T_i i.e.

$$\frac{2a}{RT_i} = b \quad \text{or} \quad T_i = \frac{2a}{Rb} \qquad ...(9)$$

Therefore if Vander Walls constants are known T_i can be calculated.

For most of the gases, the ordinary working temperature are below their temperature of inversion and so they show a cooling effect. However the Temperature of inversion for hydrogen and helium are much below the ordinary temperature and hence at these temperatures they show a heating effect. If, however, these gases are precooled below their inversion temperature before allowing to undergo a Joule-Thomson expansion, they will also saw a cooling effect.

By Joule-Thomson Effect. The cooling effect observed when a gas under pressure as allowed to expand, through a narrow orifice, into a region of low pressure, is known as Joule-Thomson effect.

In most of the gases cooling is observed but in case of hydrogen and helium and heating effect is observed.

For every gas there is a certain temperature below which a cooling effect will be observed. Such temperature is known as *inversion temperature.* For hydrogen and Helium inversion temperature is very low (–80°C and –240°C respectively). For other gases it lies above room temperature.

The cooling effect of this process can be explained because the molecules do internal work in order to over come the forces of attraction between them.

One of the process uses for cooling and liquifaction of air based on Joule-Thomson effect is Linde's process. The apparatus used for the process in shown below in Fig. 5.5.

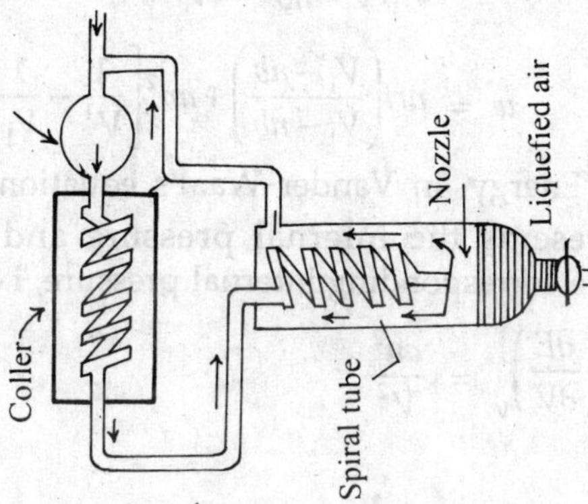

Fig. 5.5. Linde's Apparatus.

Air free from moisture and carbon dioxide is compressed to about 200 atmospheres, it is then cooled by passing through a water cooler and then allowed to expand to about 50 atmospheres through a small orifice. Because of Joule-Thomson effect the temperature falls and air gets cooled. This process is repeated till air liquifies.

5.6 By the Adiabatic Expansion of Cold Compressed Gases

Work of Expansion of Isothermal and Adiabatic Expansion of Gases

Reversible Isothermal Expansion of Real Gas Obeying Vander Waal's Equation

Work of Expansion. We have already proved that

$$dw = PdV \qquad \ldots(1)$$

or

$$w = \int_{V_1}^{V_2} PdV \qquad \ldots(2)$$

For *n* moles, the Vander Waal's equation is

$$P = \frac{nRT}{V-nb} - \frac{an^2}{V^2} \quad ...(3)$$

On substituting equation (3) in (2), we get

$$w = \int_{V_1}^{V_2} \frac{nRT}{V-nb} dV \int_{V_1}^{V_2} \frac{an^2}{V^2} dV$$

or

$$w = nrt\left(\frac{V_1 - nb}{V_1 - nb}\right) + an^2\left[\frac{1}{V^1} - \frac{1}{V_1}\right]$$

Internal Energy. In Vander Waal's equation, the factor an^2/V^2 represents the internal pressure and $(\partial E/\partial V)_T$ represents the corresponding internal pressure, i.e.,

$$\left(\frac{\partial E}{\partial V}\right)_V = \frac{an^2}{V^2}$$

or

$$dE = \frac{an^2}{V^2} dV \qquad \text{(at constant T)} \quad ...(5)$$

On integrating the above equation between suitable limits, we get

$$\int_1^2 dE = \int_{V_1}^{V_2} \frac{an^2}{V^2} dV$$

or

$$E_1 - E_1 = -an^2\left[\frac{1}{V_2} - \frac{1}{V_2}\right] \quad ...(6)$$

or

$$\Delta E = -an^2\left[\frac{1}{V_2} - \frac{1}{V_1}\right] \quad ...(7)$$

Enthalpy change : We know

$$H = E + PV$$

For the initial state, the enthalpy is

and for the final state, the enthalpy is

$$H_2 = E_2 + P_2V_2$$

Enthalpy change = $H_2 - H_1$

or $$\Delta H = (H_2 + P_2V_2) - (E_1 + P_1V_1)$$

$$\Delta H = (H_2 - E_2) + (P_2V_2 - P_1V_1)$$

or $$= \Delta E + (P_2V_2 - P_1V_1) \quad ...(8)$$

The Vander Wall's equation for *n* moles is

$$PV = \frac{nRT}{V-b} - \frac{an^2}{V^2}$$

On multiplying the above equation by *V*, we get

$$PV = \frac{nRTV}{V-nb} - \frac{an^2}{V}$$

For the initial state,

$$P_1V_1 = \frac{nRTV_1}{V_1-nb} - \frac{an^2}{V_1} \quad ...(9)$$

for the final state,

$$P_2V_2 = \frac{nRTV_2}{V_2-nb} - \frac{an^2}{V_2} \quad ...(10)$$

On substituting equations (9) and (10) in (8), we get

$$\Delta H = \Delta E + nRT\left[\frac{V_2}{V_2-nb} - \frac{V_1}{V_1-nb}\right] + an^2\left[\frac{1}{V_1} - \frac{1}{V_2}\right] \quad ...(11)$$

Here

$$\frac{V_2}{V_2-nb} - \frac{V_1}{V_1-nb} = \frac{V_2(V_1-nb) - V_1(V_2-nb)}{(V_2-nb)(V_1-nb)}$$

$$= \frac{nb(V_1-V_2)}{(V_2-nb)(V_1-nb)} = \frac{nb}{V_2-nb} - \frac{nb}{V_1-nb} \quad ...(12)$$

On substituting equation (12) in (11), we get

$$\Delta H = \Delta E + nRT\left[\frac{nb}{(V_2-nb)} - \frac{nb}{(V_1-nb)}\right] + an^2\left(\frac{1}{V_1} - \frac{1}{V_2}\right)$$

We have already proved that

$$\Delta E = -an^2\left(\frac{1}{V_1}-\frac{1}{V_2}\right) \qquad \text{[See eq. (7)]}$$

$$\therefore \qquad \Delta E = -an^2\left(\frac{1}{V_1}-\frac{1}{V_2}\right)+nRT\left[\frac{nb}{V_2-nb}-\frac{nb}{V_1-ab}\right]$$

$$+an^2\left(\frac{1}{V_1}-\frac{1}{V_2}\right)$$

$$= +n^2bRT\left[\frac{1}{V_2-nb}-\frac{1}{V_1-ab}\right]+2an^2\left[\frac{1}{V_1}-\frac{1}{V_2}\right] \qquad \ldots(13)$$

Heat Change q: According to the first law of thermodynamics.

$$q = \Delta E = w$$

We have already proved that

$$\Delta E = an^2\left[\frac{1}{V_1}-\frac{1}{V_2}\right] \qquad \text{[see eq. (7)]}$$

$$w = nRT\ln\left(\frac{V_2-nb}{V_1-nb}\right)+an^2\left(\frac{1}{V_2}-\frac{1}{V_1}\right) \qquad \text{[See eq. (4)]}$$

$$\therefore \quad q = an^2\left(\frac{1}{V_2}-\frac{1}{V_1}\right)+nRT\ln\left(\frac{V_2-nb}{V_1-nb}\right)+an^2\left(\frac{1}{V_2}-\frac{1}{V_1}\right)$$

$$= nRT\ln\left[(V_2-nb)/(V_1-nb)\right] \qquad \ldots(14)$$

Reversible Adiabatic Expansion of Real Gas Obeying Vander Wall's Equations

When reversible adiabatic expansion of real gas takes place, there is no exchange of heat between the system and the surroundings, i.e.,

$$q = 0$$

First law of thermodynamics gives,

$$\Delta E = w \quad \text{...(1)}$$

If we know ΔE, w can be calculated.

As E is a state function of T and V, it means that

$$E = f(T, V) \quad \text{...(2)}$$

$$dE = \left(\frac{\partial E}{\partial T}\right)_V dT + \left(\frac{\partial E}{\partial T}\right)_V dV \quad \text{...(3)}$$

For a Vander Wall's equation,

$$\left(P + \frac{n^2 a}{V^2}\right)(V - nb) = nRT$$

$$\left(\frac{\partial E}{\partial T}\right)_V = nC_V \text{ and} \left(\frac{\partial E}{\partial V}\right)_T = \frac{an^2}{V^2} \quad \text{...(4)}$$

On substituting equation (4) in (3), we get

$$dE = nC_V dT + \frac{an^2}{V^2} dV \quad \text{...(5)}$$

If it is assumed that C_v is independent of temperature T, the equation (5) can be intergrated between the suitable limits to yield

$$\int_1^2 dE = nC_v \int_{T_1}^{T_2} dT + an^2 \int_{V_1}^{V_2} \frac{dV}{V^2}$$

or

$$E_2 - E_1 = \Delta E = nC_v(T_2 - T_1) - an^2 \left[\frac{1}{V_2} - \frac{1}{V_1}\right]$$

or

$$\Delta E = nC_v(T_2 - T_1) - an^2 \left[\frac{1}{V_2} - \frac{1}{V_1}\right] \quad \text{...(6)}$$

When the value of ΔE calculated from equation (5) is substituted in equation (1) the value for w can be calculated.

Relation Between T and V: For a differential change in an adiabatic process, the first law may be put as

$$dE = -PdV \quad ...(7)$$

$$dE = nC_v dT + \left(\frac{\partial E}{\partial V}\right)_V dV \quad ...(8)$$

For a Vander Waal's equation, we know

$$\left(\frac{\partial E}{\partial V}\right)_V = \frac{a}{V_2} \text{ (for } n = 1 \text{ mole)} \quad ...(9)$$

and

$$P = \frac{RT}{V-b} - \frac{a}{V_2} \text{(for } n = 1 \text{ mole)} \quad ...(10)$$

$$dE = C_V dT + \frac{a}{V^2} dV = -\left[\frac{RT}{V-b} - \frac{a}{V^2}\right] dV$$

or

$$C_V dT = -\left(\frac{RT}{V-b}\right) dV$$

or

$$\frac{C_V}{R}\frac{dT}{T} = \frac{dV}{V-b}$$

On integrating the above equation between the suitable limits, we get

$$\frac{C_V}{R}\int_{T_1}^{T_2} d\ln T = -\int_{V_1}^{V_2}\frac{dV}{V-b}$$

or

$$\frac{C_V}{R}\ln\frac{T_2}{T_1} = 1n\frac{(V_1 - b)}{(V_2 - b)}$$

or $$(T_2)^{C_V/R}(V_2 - b) = (T_1)^{C_V/R}(V_1 - b) \quad ...(11)$$

or $$(T)^{C_V/R}(V - b) = \text{constant} \quad ...(12)$$

Equation (10) can be simplified to get

$$T_2 = T_1\left(\frac{V_1 - b}{V_2 - B}\right)^{R/C_V} \quad ...(13)$$

SOLVED EXAMPLES

Example 1. *Express the following quantities in the alternative forms specified:*

(a) *2000 kW in MW* (b) *002 m^3 in dm^3*

(c) *10 bar in kPA* (d) *1000 pF in μF*

(e) *10 MJ in J* (f) *1 gm/cm^3 in kg/m^3*

Solution.

$$\text{(a) } 2000 \text{ kW} = \frac{2000\times10^3}{10^6} = 2\,MW$$

$$\text{(b) } .002\ n^3 = 0.02\times\left(\frac{10}{1}\right)^2 = 2\text{ dm}^3$$

$$\text{(c) } 10 \text{ bar} = 10\times\frac{10^5}{10^3} = 1000\ kPa$$

$$\text{(d) } 1000\ pF = 1000\times\frac{10^{-12}}{10^{-6}} = .001\mu F$$

$$\text{(e) } 10 \text{ MJ} = 10\times10^6 = 1\times10^7\,J$$

$$\text{(f) } 1\text{ g/cm}^3 = \frac{1\times10^{-3}}{(1/100)^3} - 1000\text{ kg/m}^3.$$

Example 2. *What will be the volume occupied by 3 moles of an ideal gas at 150 kPa and 300 K. (R = 8.314 J/mol K)*

Solution. $PV = nRT$ (Gas equation)

$$V = \frac{nRT}{P}$$

$$= \frac{3\times8.314\times300}{150\times1000}$$

$$= 0.04988\text{ m}^3.$$

[Note that the answer has been obtained in m^3 by converting pressure directly to P_a (or N/m^2), since R and T were already in base units.]

Example 3. *Calculate the minimum work required to form the surface associated with a water mist having a particle radius of 10 μm? 1 dm³ of water at 50°C is to be turned into mist at 50°C. Surface tension of water at 50°C is 63 mN/m.*

Solution.
$$A = \frac{3V}{r}$$
$$= \frac{3\times10^{-3}}{10\times10^{-6}}$$
$$= 300\ \text{m}^2$$
$$W = \int_0^A \gamma dA = -\gamma A$$
$$= -63\times10^{-3}\times300$$
$$= -18.9\ \text{J}$$

Example 4. *A car is moving down a hill with a velocity of 30 m/s. The driver sees a red light at the bottom of the hill when he is at a vertical height 30 m above the hill. If the weight of car and driver is 1350 kg now much energy as heat must be dissipated by the brakes [wind and other frictional effects are negligible].*

Solution. $\Delta E = Q - W$

ΔE is made up of ΔE_k (K.E.) and ΔE_p (P.E.)

$$\Delta E_k = \frac{m\Delta u^2}{2} \text{ and } \Delta E_p = mg\Delta h$$

Since in this case W = 0 [∵ no moving surface force]

$$\therefore \quad Q = \frac{m\Delta u^2}{2} + mg\Delta h$$
$$= \frac{1350\times(0-30^2)}{2} + 1350\times9.8066(0-30)$$
$$= -\frac{1215000}{2} - 397167$$
$$= -1004667\ \text{J or } -10004.667\ \text{kJ},$$

Example 5. *One mole of a diatomic gas is isobarically heated from 27°C to 87°C. Calculate the increase of internal energy and the external work done.*

Solution. $$\Delta E = C_V(T_2 - T_1)$$

$$= \frac{5}{2}R(T_2 - T_1)$$

$$= \frac{5}{2} \times 8.37 \times 60$$

$$= 1.256 \times 10^3 \text{J}.$$

$$W = P(V_2 - V_1)$$

$$= R(T_2 - T_1)$$

$$= 8.37 \times 60$$

$$= 502.2 \text{ J}.$$

Example 6. *14 gms of nitrogen are heated isochorically when the pressure increases from* 2×10^5 N/m^2 *to* 2.8×10^5 N/m^2. *Calculate the change in internal energy. Initial temperature is 27°C,* $R = 8.4$ *J/mole°K.*

Solution. $$\frac{P_1}{T_1} = \frac{P_2}{T_2} \qquad \text{[Isochoric process]}$$

$$\therefore \quad T_2 = 300 \times \frac{2.8}{2} \times \frac{10^5}{10^5}$$

$$= 420°\text{K}$$

Change in internal Energy $= C_V \dfrac{m}{M}(T_2 - T_1)$

$$= \frac{5}{2}R \times \frac{14}{28} \times 120$$

$$= 1260 \text{ J}.$$

Example 7. *2 gms of hydrogen at 127°C are adiabatically expanded so that final pressure is half the initial pressure. Calculate the final temperature and the work done. g = 1.4, = 8.4 J/mole°K.*

Solution. We know, for an adiabatic process

$$\frac{T_1^{\gamma-1}}{P_1^{\gamma}} = \frac{T_2^{\gamma}}{P_2^{\gamma-1}}$$

$$\therefore \quad T_2 = T_1\left(\frac{P_2}{P_1}\right)^{\frac{\gamma-1}{\gamma}}$$

$$= 400 \times \left(\frac{1}{2}\right)^{\frac{.4}{1.4}}$$

$$= 328°K$$

$$W = n \cdot \frac{R}{\gamma - 1}(T_1 - T_2)$$

$$= 1 \times \frac{8.4}{.4}(400 - 328)$$

$$= 1512 \text{ J.}$$

Example 8. *At 500°C and between pressure 0-60 atm. μ j.t. for nitrogen can be represented by*

$$\mu j.t. = 0.0140 - 2.50 \times 10^{-4}P$$

Find the expected temperature drop when the gas is expanded from 60 to 20 atmosphere pressure.

Solution. By definition,

$$\mu \text{j.t.} = \left(\frac{\partial T}{\partial P}\right)_H$$

$$\therefore \quad dT = \mu_{J.T.} dP$$

$$= (.0140 - 2.5 \times 10^{-4}P)dP$$

Integrating with in limits

$$\Delta T = 0.140 / P\Big|_{60}^{20} - 2.5 \times 10^{-4} \frac{P_2}{2}\Big|_{60}^{20}$$

$$= 0.140 \times (-40) - \frac{2.5 \times 10^{-4}}{2}(20^2 - 60^2)$$

$$= -.560 + .4$$

$$= -.16°C.$$

Example 9. *Two moles of a Vander Walls gas at 27°C expands isothermally and reversibly. The initial volume is 20 l and the final volume becomes 40 l. Calculate Q, W, ΔE and ΔH. Given that a = 5.49 l^2-atm. mole^{-2}, b =.0641 mole^{-1}.*

Solution. In case of Vander Walls gas

$$W = nRT \ln\left(\frac{V_2 - b}{V_1 - b}\right) + a\left(\frac{1}{V_2} - \frac{1}{V_1}\right)$$

$$= 2 \times 0.82 \times 300 \times 2.303 \log\left(\frac{40-.064}{20-.064}\right)$$

$$+ 5.49\left(\frac{1}{40} - \frac{1}{20}\right)$$

$$= 226.5 \text{ litres-atm.}$$

$$\Delta E = -a\left(\frac{1}{V_2} - \frac{1}{V_1}\right)$$

$$= -5.49\left(\frac{1}{40} - \frac{1}{20}\right)$$

$$= -.135 \text{ litres atm.}$$

$$Q = nRT \ln \frac{V_2 - b}{V_1 - b}$$

$$= 2 \times 0.82 \times 300 \times 2.303 \times \log\left(\frac{40-.064}{20-.064}\right)$$

$$= 226.63 \text{ litre-atm}$$

$$\Delta H = nRTb\left[\frac{1}{V_2 - b} - \frac{1}{V_1 - b}\right] - 2a\left[\frac{1}{V_2} - \frac{1}{V_1}\right]$$

$$= 2 \times 0.82 \times 300 \times 0.64\left[\frac{1}{40-.064} - \frac{1}{20-.064}\right]$$

$$-2 \times 5.49\left[\frac{1}{40} - \frac{1}{20}\right]$$

$$= -.92 \text{ litre-atm.}$$

Example 10. *An ideal gas whose equation of state is $PV = nRT$ and whose internal energy is a function of T' only is used as the working substance in a car not cycle. Show that the ideal gas temperature T' is equal to the thermodynamic temperature T.*

Solution.

According to first law of thermodynamics

$$q = \Delta E + w$$

For an isothermal process (since T' is constant) $\Delta E = 0$,

$$\therefore \quad Q_1 = nRT_1' \ln \frac{V_b}{V_c}$$

Similarly

$$Q_2 = nRT_2' \ln \frac{V_c}{V_d}$$

But for any reversible adiabatic process

$$TV^{\gamma-1} = \text{constant}$$

$$\therefore \quad \frac{T_1'}{T_2'} = \left(\frac{Vc}{V_b}\right)^{\gamma-1} = \left(\frac{V_d}{V_a}\right)^{\gamma-1}$$

Hence

$$\left(\frac{V_b}{V_a}\right) = \left(\frac{V_c}{V_d}\right)$$

or

$$\frac{Q_1}{Q_2} = \frac{nRT_1' \ln\left(\frac{V_b}{V_a}\right)}{nRT_2' \ln\left(\frac{V_c}{V_a}\right)} = \frac{T_1'}{T_2'}$$

Thus

$$\frac{T_1}{T_2} = \frac{T_1'}{T_2'} \qquad \left[\because \frac{Q_1}{Q_2} = \frac{T_1}{T_2}\right]$$

So from above it follows that

$$T = T'$$

Hence the ideal gas temperature T' is equal to the thermodynamic temperature T.

By the Adiabatic Expansion of Cold Compressed Gases

During expansion work is done by the gas. This work is done against external pressure. If this expansion is carried out under adiabatic conditions (i.e. $\Delta H = 0$), the temperature falls. If a highly compressed gas is suddenly allowed to expand adiabatically, it has to do a large amount of external work and so the cooling effect will be large.

The principle was applied to the liquifaction of gases by claude, who used it to produce liquid air cailletet used it for liquifying air and many other gases like O_2, C_2H_2, CO etc. The apparatus used by him consisted of a thick glass tube in which the gas to be cooled was taken over mercury. The pressure was then raised by pumping more mercury. The gas could be alternately compressed and expanded rapidly.

Claud's method for liquifaction of air is a modified method based on this principle. The process is shown diagrammatically in Fig. 5.6.

Air free from carbon dioxide and moisture is compressed to about 40 atmospheres pressure and then passed through XYZ pipe.

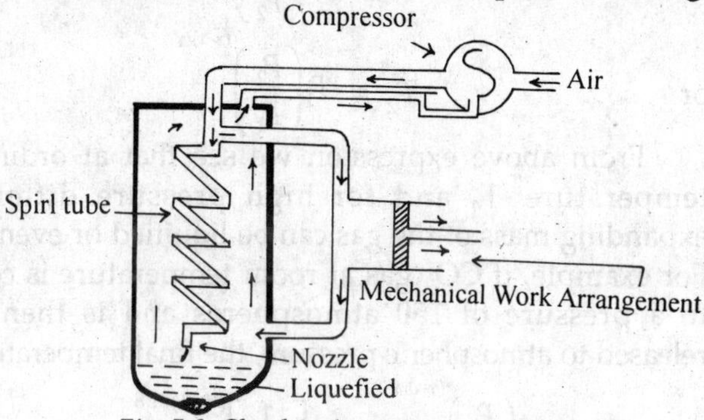

Fig. 5.6. Claude's Apparatus.

This pipe splits into two parts at Z. Some of the air goes to D directly where as some enters C. C is fitted with air tight piston. The air entering it expands adiabatically and so its cools. This cooled air goes up the chamber D and helps in cooling the air passing down through DZ. This chilled air expands out of Jet and it is further cooled. This is continued to the air condenses into liquid and is collected at the bottom as shown.

The cooling effect produced by this method is quite large as can be seen from the following discussion:—

Let P is be the pressure and T be the temperature of a perfect gas, then during adiabatic process

$$PV\gamma = \text{constant}$$

where, γ is the ratio of two specific heats (i.e. $\gamma = \frac{c_p}{c_u}$)

Using gas equation $PV = RT$, we must have

$$TP(1-\gamma)/\gamma = \text{constant}$$

Let T_1 and P_1 are the initial temperature and pressure of a given mass of the gas, and T_2 and P_2 are its final temperature and pressure, then we have and

$$T_1P_1(1-g)/g = T_2P_2^{(1-\gamma)/\gamma}$$

or

the final temperature $T_2 = T_1\left(\dfrac{P_1}{P_2}\right)^{(1-\gamma)/\gamma}$ ...(*i*)

or $$T_2 = T_1\left(\frac{P_2}{P_1}\right)^{(\gamma-1)/\gamma} \quad ...(ii)$$

From above expression we see that at ordinary initial temperature T_1 and for high pressure difference, the expanding mass of the gas can be liquified or even solidified. For example, if CO_2 gas at room temperature is compressed to a pressure of 150 atmospheres and is then suddenly released to atmospheric pressure, the final temperature (T_2) is

$$T_2 = T_1\left(\frac{P_2}{P_1}\right)^{(\gamma-1)/\gamma} = 300\left[\frac{1}{150}\right]^{(½-1)/1.8} = 96K.$$

This temperature is so low, that the expanding mass of CO_2 gets solidified. The solidified mass is dry ice.

The machine based on the above principle is called *gas expansion machine*. The working of such a machine is represented in Fig. 5.7.

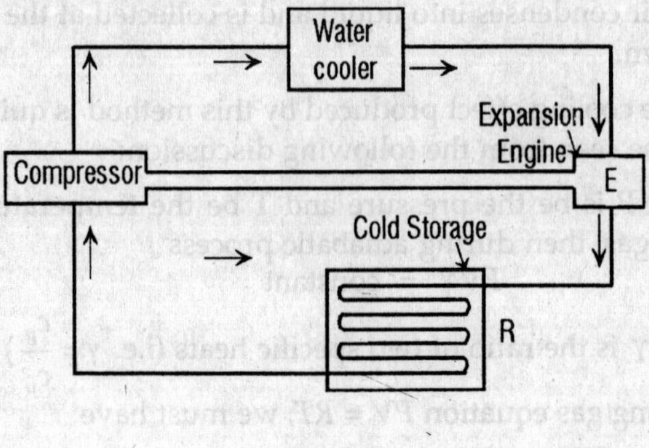

Fig. 5.7.

It is made up of the following four main parts:

1. Compressor P that is driven by an electric motor.
2. Water cooler C.
3. An expansion engine E.
4. A reservoir or cold storage space R.

The gas or air is first compressed in the compressor P adiabatically. The compressed gas is then passed through a water cooler C, where it gets cooled to a suitable temperature at constant pressure. This cooled gas is then passed to expansion engine where it undergoes adiabatic expansion to a pressure equal to that of the intake of the compressor and cooled to low temperature. This cold gas is then further passed to the reservoir or the cold storage space where it absorbs more heat at constant pressure and finally returns to the compressor so that a fresh cycle of operation starts. The power for the compressor is partly obtained from expansion engine and partly from the electric motor.

Coefficient of performance of a gas expansion machine. The working cycle of the machine is represented on a *P-V* diagram in Fig. 5.8. The gas at low pressure P_1 (generally atmospheric pressure) enters the compressor at point A where the temperature of the gas is *Ta*. Here the gas is compressed adiabatically to pressure P_2, high compared to P_1, temperature of the gas being T_b (point B). The compressed

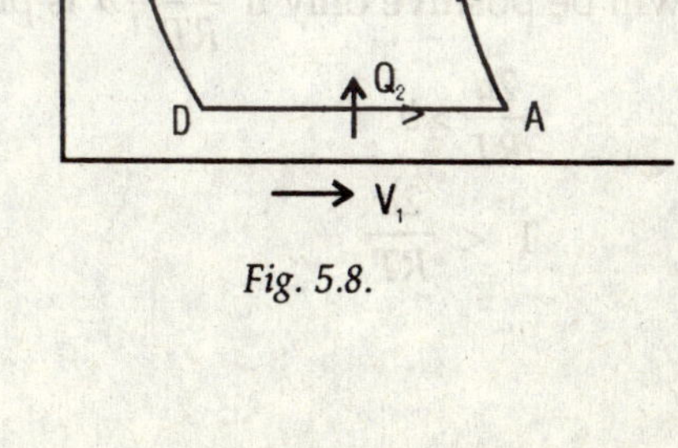

Fig. 5.8.

gas is cooled at this pressure through the cooler to the temperature T_c of the coolant (point C). The gas is then passed to expansion engine where it is permitted to expand adiabatically along path CD to the point D where the pressure and temperature of the gas are P_1 and T_d respectively. This cooled gas is then passed through the reservoir R from which it absorbs more heat at constant pressure along path DA. As the gas reaches the state A, the cycle is completed and a fresh one starts.

Let the quantity of the working gas be 1 gm. molecule.

The quantity of heat given to the coolant.

Air free moisture and carbon dioxide is compressed to about 200 atmospheres, it is then cooled by passing through a water cooler and then allowed to expand to about 50 atmospheres through a small orifice. Because of Joule-Thomson effect the temperature falls and air gets cooled. This process is repeated till the air liquifies.

It should be clearly understood that only *a gas below its inversion temperature shows cooling effect.*

A heating will take place if the temperature of the gas is above its critical temperature.

We know that the fall in temperature, dT, for a gas obeying Vander Waals equation is given by

$$dT = \frac{1}{C_P}\left(\frac{2a}{RT} - b\right)dP \qquad \ldots(i)$$

where dP is the difference in pressure on the two sides of the porous plug.

From above expression, it is clear that

(i) dT will be positive only if $\frac{2a}{RT} - b$ is positive i.e.,

$$\frac{2a}{RT} > b$$

or

$$T < \frac{2a}{RT}$$

Therefore, the cooling will take place only if the temperature of the gas is less than 2a/Rb.

(ii) dT will be zero if $\frac{2a}{RT} - b = 0$, i.e. $T = \frac{2a}{Rb}$

Therefore, the fall in temperature will be zero if the temperature of the gas is equal to 2a/Rb. This is called the *temperature of inversion* and is represented by T_i. Thus the temperature of inversion,

$$T_i = \left(\frac{2a}{Rb}\right) \qquad \ldots(ii)$$

(iii) *d*T will be negative if $\left(\frac{2a}{Rb} - b\right)$ is negative

i.e., $$b > \frac{2a}{RT}$$

or $$T > \frac{2a}{Rb}$$

or $$T > T_i$$

Therefore, the heating will take place if the temperature of the gas is greater than the temperature of inversion.

Inversion curve. Supposing that a series of Joule-Thomson experiments are performed on the same gas in the following way. The initial pressure and temperature P_1 and T_1 (on the high pressure side of the plug) are chosen arbitrarily and the pressure P_2 on the other sides of the plug is fixed at any desired value less than P_1 and then the temperature T_2 of the escaping gas is measured. Keeping P_1 and T_1, constant P_2 is changed to another value and corresponding T_2 is measured. This is done many a number of times for different values of P_2 and the different values of the final temperature T_2 in each case measured. Now a graph is plotted between various values of P_2 and the corresponding values of T_2, a curve is obtained as shown in Fig. 5.9: One point on the curve is (P_1, T_1) and the others, being indicated by, a, b, c, d..., represent various P_2's and T_2's. Since in the Joule-Thomson expansion, the initial and final enthalpies are equal, the enthalpy at all points a, b, c is the same and the curve drawn through these points represents a constant enthalpy (isenthalpic curve). An isenthalpic curve is the locus of all points representing equilibrium state of the same enthalpy.

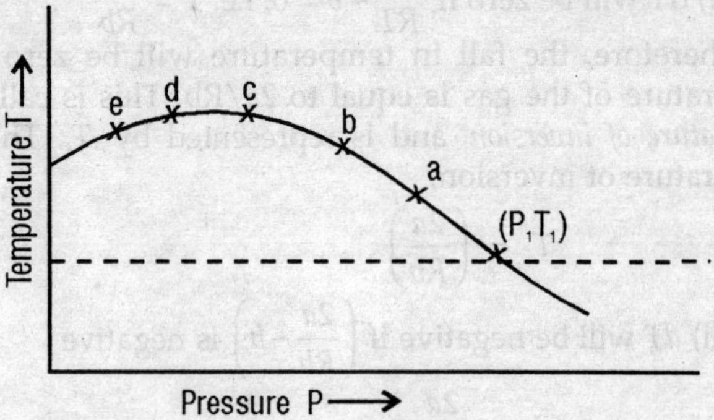

Fig. 5.9

Now the temperature T_1 on the high pressure side of the plug is changed to another value keeping P_1 the, P_2 is again constant changed and corresponding T_2's measured. By performing other series of experiments with the initial temperature T_1 same for one series; but different for different series, a family of curves corresponding to different values of enthalpy H are obtained as showing in Fig. 5.10. If the temperature is not very high, such curves pass through a maximum known as the *inversion point*. The locus of all the inversion points is called the *inversion curve* (shown in Fig. 5.10). Since towards the left of the maximum of either curve of constant enthalpy (isenthalpic curve) the gas shows cooling effect, the region inside inversion curve is the *cooling region*, for in this region gas shows *cooling effect*. On the other hand, the region outside the inversion curve is called the region of heating because in this region the gas shows heating effect. It can be seen that at any given pressure less than the extreme value, there are two inversion temperatures. One is the "ordinary" inversion temperature (higher value) at which the heating effect gives place to cooling; and the other a much lower temperature at which cooling gives place again to heating. The ratio of these two inversion temperatures for a Vander Waal's gas is approximately 9. To produce cooling effect, the gas must be precooled below a temperature where the inversion curve cuts the temperature axis on *P-T* diagram, i.e., the maximum inversion temperature. For most of the gases ordinary temperature is already below the maximum inversion temperature, and so no precooling is essential in these cases. However, for hydrogen and helium, even the maximum inversion temperatures are much below the ordinary temperatures and hence at ordinary temperatures they show a heating effect. These gases must, therefore, be precooled upto – 80°C and –243°C respectively and then they will also show a cooling effect.

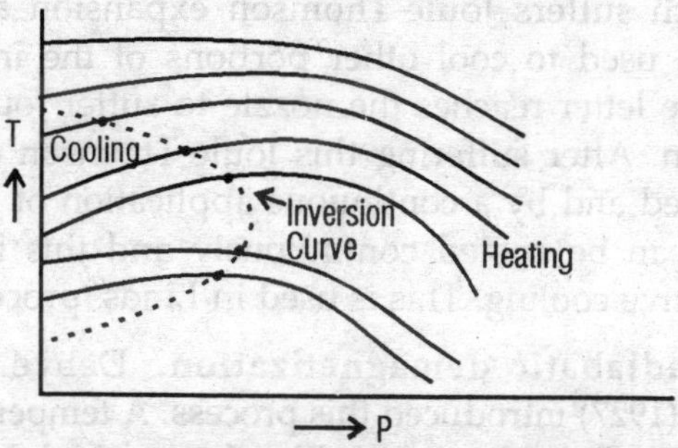

Fig. 5.10

It must be carefully noted that the inversion temperature is not constant for a given gas but varies with the initial pressure of the gas.

Example 7. *For one mole of hydrogen, the Vander Waal's constants are*

$$a = 0.245 \frac{litre^2 \times atoms}{mole^2}$$

and

$$b = 2.67 \times 10^{-2} \; litre \; mole^{-1}.$$

Calculate its temperature of inversion. Given that R = 2 cal. mole⁻¹K⁻¹ and 1 atmosphere = 106 dynes/cm³.

$$T_i = \frac{2a}{Rb}$$

Here

$$a = 0.245 \text{ litre}^2 \text{ atoms/mole}^2$$

$$= 0.245 \times 10^6 \times 10^6 \text{ cm}^4 \text{ dynes/mole}^2$$

$$b = 2.67 \times 10^{-2} \text{ litre/mole}$$

$$= 26.7 \text{ cm}^3\text{/mole}$$

and $$R = 2 \text{ cal/mole K}$$

$$= 2 \times 4.2 \times 10^7 \text{ ergs/mole K}$$

Hence

$$T_i = \frac{2 \times 0.245 \times 10^{12}}{2 \times 4.2 \times 10^7 \times 26.7}$$

$$= 220\text{K}.$$

Regenerative cooling. In this process, a portion of the gas which suffers Joule-Thomson expansion and because cooled is used to cool other portions of the incoming gas before the letter reaches the nozzle to suffer Joule-Thomson expansion. After suffering this Joule-Thomson expansion it gets cooled and by a continuous application of this process, the gas can be cooled continuously and this is known as regenerative cooling. This is used in Linds' process.

By adiabatic demagnetization. Debye (1926) and Gianque (1927) introduced this process. A temperature as low as 1K (-272°C) can be attained by this method. It is based on the fact that when a paramagnetic substance is magnetised it results in the increase of orderliness of the system and the reverse of it i.e., demagnetization results in disorder.

During magnetization work is done on the system and during demagnetization work is done by the system and for this purpose it draws energy from with in the system itself. This results in the cooling of the substance and the surroundings.

If we work on paramagnetic salts of gadolinium, Iron etc., temperatures as low as 05°K can be obtained Dehye, Gianque could produce a temperature of .25°K by working on gadolinium sulphate. De Hass wierswe working on a mixture of potash alum and chrome alum was able to attain a temperature of .003K (1946).

A considerable cooling can thus the produced by using strong magnetic field and low initial temperature.

This is in accordance with curie law, which states, the magnetic susceptibility of a substance is inversely proportional to the absolute temperature i.e.,

$$\chi \alpha \frac{1}{T} \text{ or } \chi = \frac{C}{T}$$

We can determine the final temperature by measuring the value of χ (magnetic susceptibility).

With the help of this process temperatures as low as of the order of 10^{-4} can be attained. The precise measurement of such a low temp is quite difficult and the temperature is actually calculated with the help of curie law by measuring the magnetic susceptibility (χ). The temperature so calculated is known as *curie temperature* (T^*)

$$T^* = \frac{C}{\chi}$$

Curie Temperature (T^*) is also called Magnetic temperature.

Though temperature as low as 0.01 0K have been produced but the attainment of absolute zero (0K) is an impossibility.

Example *Calculate the cooling produced by adiabatic demagnetisation of a paramagnetic salt as the field is reduced from 5000 Oersted to zero at initial of temperature 2K. Given Curie constant per gm. mole. per c.c. = .06C. G.S. units and CH = .042 joule gm*$^{-1}$ *deg*$^{-1}$.

Solution. The final temperature is given by

$$T_2 = T_1 \left[1 - \frac{k}{C_H} \left(\frac{H}{T_1} \right)^2 \right]^{1/2}$$

Given

$$T_1 = 2\text{K}, CH = 0.42 \text{ joule gm}^{-1} \text{ deg}^{-1}$$

$$= .42 \times 10^7 \text{ erg gm}^{-1} \text{deg}^{-1}$$

and Curie constant, k =.06

$$T_2 = 2\left[1 - \frac{0.06}{0.42 \times 10^7}\left(\frac{5000}{2}\right)^2\right]^{1/2}$$

$$= 1.82 \text{ K}$$

Therefore cooling produced = $T_1 - T_2$ = (2-1.82) K

= 0.18 K.

5.4 Boyles Temperature and Critical Temperature of a Gas

All gases are composed of molecules and can be liquified by increase of pressure and/or decrease of temperature. However in the luquifaction of gases temperature plays more important part than pressure. It is possible to liquify all gases at atmospheric pressure but some of them can not be liquified at room temperature.

Thus, for every gas there is a temperature above which it can not be liquified by increase of pressure. This is known as critical temperature.

Critical temperature is the minimum temperature which must be attained before a gas can be liquified by the application of pressure.

Each gas possesses a definite critical temperature. For some gases these as follow:

Gas	*Critical Temperature (Tc)*
Helium	5.20A
Hydrogen	33.20A
Oxygen	154.28A
Nitrogen	125.97A
CO_2	304.15A
Cl_2	417.00A
NH_3	405.50A

6

Partial Molar Properties of Gases

6.1 INTRODUCTION

To define the state of a system, any pair of quantities P, V, T, S, E, H, A, G may be used. We know that E, H, S, A, G are state functions so their differentials are exact.

We have the following equations for some of these functions:

$$dE = dQ - PdV \quad ...(1)$$

$$dH = dQ + VdP \quad ...(2)$$

From First law of thermodynamics

and

$$dE = Tds - PdV \quad ...(3)$$

$$dA = TdS + PdV \quad ...(4)$$

$$dA = -SdT - PdV \quad ...(5)$$

$$dG = -SdT + VdP \quad ...(6)$$

From Second law of thermodynamics

These equations give us independent variables which may be used for each of the four functions:

$$E = f(S,V)$$

$$H = f(S,P)$$

$$A = f(T,V)$$

$$G = f(T,P)$$

Since dF, dH, dA and dG are exact differentials, therefore

$$\left(\frac{\partial E}{\partial S}\right)_V = T \text{ and } \left(\frac{\partial E}{\partial V}\right)_S = -P \quad \ldots(7)$$

$$\left(\frac{\partial H}{\partial S}\right)_P = T, \left(\frac{\partial H}{\partial P}\right)_S = V \quad \ldots(8)$$

$$\left(\frac{\partial A}{\partial T}\right)_V = S, \left(\frac{\partial A}{\partial V}\right)_T = -P \quad \ldots(9)$$

$$\left(\frac{\partial G}{\partial T}\right)_P = -S, \left(\frac{\partial G}{\partial P}\right)_T = V \quad \ldots(10)$$

From these equations we can see the change in one dependent variable with a change in one independent variable, keeping the appropriate variable constant.

Since these are exact differentials so Euler's theorem is applicable. Hence we get the following relations known as Maxwells relations.

Applying Euler's reciprocity theorem to equations 3, 4, 5, 6 we get

$$\left(\frac{\partial T}{\partial V}\right)_S = -\left(\frac{\partial P}{\partial S}\right)_V \quad \ldots(11)$$

$$\left(\frac{\partial T}{\partial P}\right)_S = \left(\frac{\partial V}{\partial S}\right)_P \quad \ldots(12)$$

$$\left(\frac{\partial S}{\partial V}\right)_T = \left(\frac{\partial P}{\partial T}\right)_V \quad \ldots(13)$$

and

$$\left(\frac{\partial S}{\partial P}\right)_T = -\left(\frac{\partial V}{\partial T}\right)_P \quad \ldots(14)$$

By reversing the equations (11) and (12), we get

$$\left(\frac{\partial S}{\partial P}\right)_V = -\left(\frac{\partial V}{\partial T}\right)_S$$

and $$\left(\frac{\partial S}{\partial V}\right)_P = \left(\frac{\partial P}{\partial T}\right)_S$$

All these are important as they relate the rate of change of a particular quantity to P, V and T.

6.2 MAXWELLS RELATIONS

We know the following relations:

$$dE = TdS - PdV$$

$$dH = TdS + VdP$$

$$dA = -SdT - PdV$$

$$dG = -SdT + VdP$$

These relations can also be written as under:

(*i*) **Relation I.** Taking T and V an independent variables we have

$$\left(\frac{\partial S}{\partial V}\right)_T = \left(\frac{\partial P}{\partial T}\right)_V \quad \text{[equation 13]}$$

But

$$dS = \frac{dH}{T}$$

$$\therefore \quad \left(\frac{\partial H}{\partial V}\right)_T = T\left(\frac{\partial P}{\partial T}\right)_V \quad ...(15)$$

(*ii*) **Relation II.** Taking T and P as independent variables,

$$\left(\frac{\partial S}{\partial P}\right)_T = \left(\frac{\partial V}{\partial T}\right)_P \quad \text{[equation 14]}$$

But

$$dS = \frac{dH}{T}$$

$$\therefore \quad \left(\frac{\partial H}{\partial P}\right)_T = -\left(\frac{\partial V}{\partial T}\right)_P \quad ...(16)$$

(iii) **Relation III.** Taking S,V as independent variables,

$$\left(\frac{\partial T}{\partial V}\right)_S = \left(\frac{\partial V}{\partial S}\right)_V \qquad \text{[equation 11]}$$

But

$$dS = \frac{dH}{T}$$

$$\therefore \quad \left(\frac{\partial T}{\partial P}\right)_S = -T\left(\frac{\partial P}{\partial H}\right)_V \qquad ...(17)$$

(iv) **Relation IV.** Taking S,V as independent

$$\left(\frac{\partial T}{\partial P}\right)_S = \left(\frac{\partial V}{\partial S}\right)_P \qquad \text{[equation 12]}$$

But

$$dS = \frac{dH}{T}$$

$$\therefore \quad \left(\frac{\partial T}{\partial P}\right)_S = \left(\frac{\partial V}{\partial H}\right)_P \qquad ...(18)$$

(v) **Relation V.** Taking P,V as independent we can derive the following relation by making suitable substitutions.

$$\left(\frac{\partial T}{\partial P}\right)_S\left(\frac{\partial S}{\partial V}\right)_P - \left(\frac{\partial T}{\partial V}\right)_P\left(\frac{\partial S}{\partial V}\right)_V = 1 \qquad ...(19)$$

(vi) **Relation VI**

and $$\left(\frac{\partial P}{\partial T}\right)_S\left(\frac{\partial V}{\partial S}\right)_T - \left(\frac{\partial P}{\partial S}\right)_T\left(\frac{\partial V}{\partial T}\right)_S = 1 \qquad ...(20)$$

The relations 15 to 20 are important relations and are used in solving problems.

6.3 APPLICATIONS OF MAXWELL'S RELATIONS

(i) *To derive relation between C_p and C_V an ideal gas,*

We have $$C_p = \left(\frac{\partial H}{\partial T}\right)_P$$

$$= T\left(\frac{\partial S}{\partial T}\right)_V \qquad \left[dS = \frac{dH}{T}\right]$$

and

$$C_v = \left(\frac{\partial H}{\partial T}\right)_V = T\left(\frac{\partial S}{\partial T}\right)_V$$

Taking S as a function of T and V,

$$dS = \left(\frac{\partial S}{\partial T}\right)_V dT + \left(\frac{\partial S}{\partial V}\right)_T dV$$

or

$$\left(\frac{\partial S}{\partial T}\right)_P = \left(\frac{\partial S}{\partial T}\right)_V + \left(\frac{\partial S}{\partial V}\right)_T \left(\frac{\partial V}{\partial T}\right)_P$$

We also know that

$$\left(\frac{\partial S}{\partial V}\right)_T = \left(\frac{\partial P}{\partial T}\right)_V \qquad \text{[Maxwell relation]}$$

So we get, by substitution,

$$\left(\frac{\partial S}{\partial T}\right)_P = \left(\frac{\partial S}{\partial T}\right)_V + \left(\frac{\partial P}{\partial T}\right)_V \left(\frac{\partial V}{\partial T}\right)_P \qquad \ldots(21)$$

Multiply by T,

$$T = \left(\frac{\partial S}{\partial T}\right)_P = T\left(\frac{\partial S}{\partial T}\right)_V + T\left(\frac{\partial P}{\partial T}\right)_V \left(\frac{\partial V}{\partial T}\right)_P \ldots(22)$$

Hence

$$C_p = C_V + T\left(\frac{\partial P}{\partial T}\right)_V \left(\frac{\partial V}{\partial T}\right)_P$$

or

$$C_p - C_V = T\left(\frac{\partial P}{\partial T}\right)_V \left(\frac{\partial V}{\partial T}\right)_P \qquad \ldots(23)$$

(*ii*) *To prove that for an ideal gas* $C_p - C_V = R$.

For an ideal gas, we have

$$PV = RT \qquad \text{(one mole)}$$

$$\left(\frac{\partial P}{\partial T}\right)_V = \frac{R}{V} \text{ and } \left(\frac{\partial V}{\partial T}\right)_P = \frac{R}{P}$$

But $$C_p - C_V = T\left(\frac{\partial P}{\partial T}\right)_V\left(\frac{\partial V}{\partial T}\right)_V \qquad \text{[use eq. 23]}$$

$$\therefore \quad C_p - C_V = T\times\frac{R}{V}\times\frac{R}{P}$$

$$= \frac{TR^2}{VP} = \frac{TR^2}{RT} = R$$

Hence $C_V - C_V = R.$

(iii) To derive a relation between C_p and C_V in case of a Vander Waal's gas.

For one mole of such a gas,

$$\left(P+\frac{a}{V^2}\right)(V-b) = RT$$

or

$$\left(P+\frac{a}{V^2}\right) = \frac{RT}{V-b}$$

Differentiating, we get

$$\left(\frac{\partial P}{\partial T}\right)_V = \frac{R}{V-b}$$

Again differentiating *w,r,t,* T at constant P, we get

$$-\frac{2a}{V^3}\left(\frac{\partial V}{\partial T}\right)_P = \frac{R}{V-b} - \frac{RT}{(V-b)^2}\left(\frac{\partial V}{\partial T}\right)_P$$

or

$$\left(\frac{\partial V}{\partial T}\right)_P\left[\frac{RT}{(V-b)^2} - \frac{2a}{V^3}\right] = \frac{R}{V-b}$$

We have already derived.

$C_p - C_V$ = Substituting, we get $\left(\frac{\partial P}{\partial T}\right)_V\left(\frac{\partial V}{\partial T}\right)_P$ [eq. 23]

$$C_p – C_V = \frac{T\left(\frac{R}{V-b}\right)\left(\frac{R}{V-b}\right)}{\left[\left(\frac{RT}{V-b}\right)^2 \frac{2a}{V^3}\right]}$$

$$= \frac{RT(V-b)^2}{RT(V-b)^2}\left[\frac{R}{1-\frac{2a(V-b)^2}{V^3RT}}\right]$$

b can be neglected as compared to V

$$\therefore \qquad C_p - C_V = \frac{R}{1-\frac{2a}{VRT}}$$

$$= R\left(1-\frac{2a}{VRT}\right)^{-1}$$

Applying binomial theorem, we get

$$C_p – C_v = R\left[1+\frac{2a}{RVT}\right]$$

[∵ a is small neglect higher terms]

(iv) *Energy Equation*. The energy equations for ideal gas and gas obeying Vander Waal's gas equation have already been derived. These can also be derived as under:

For an infinitesimal change in a reversible process. ΔE is given by

$$dE = TdS - PdV$$

But

$$TdS = C_V dT + T\left(\frac{\partial P}{\partial T}\right)_V dV$$

From these two equations, we get

$$dE = C_v dT + \left[T\left(\frac{\partial P}{\partial T}\right)_V - P\right]dV \qquad ...(24)$$

Since E is a function of T and V.

So $$dE = \left(\frac{\partial E}{\partial T}\right)_V dT + \left(\frac{\partial E}{\partial T}\right)_V dV \qquad ...(25)$$

Comparing (24) and (25)

$$\left(\frac{\partial E}{\partial V}\right)_T dV = \left[T\left(\frac{\partial P}{\partial T}\right)_V - P\right] dV$$

or

$$\left(\frac{\partial E}{\partial V}\right)_T = \left[T\left(\frac{\partial P}{\partial T}\right)_V - P\right] \qquad ...(26)$$

This is known as energy equation. Internal energy of any system, whose equation of state is known, can be calculated with the help of this equation.

(i) *Ideal gas.* We know for ideal gas,

$$P = \frac{nRT}{V} \qquad ...(27)$$

or $$\left(\frac{\partial P}{\partial T}\right)_V = \frac{nR}{V} \qquad ...(28)$$

From (26) and (28), we get

$$\left(\frac{\partial E}{\partial V}\right)_T = \left[T\frac{nR}{V} - P\right]$$

or

$$\left(\frac{\partial E}{\partial V}\right)_T = P - P = 0$$

So the internal energy of an ideal gas is independent of its volume at constant temperature.

(ii) For a gas obeying Vander Waals equation

We have

$$\left(P + \frac{a}{V^2}\right)(V - b) = RT$$

$$P = \frac{RT}{V-b} - \frac{a}{V^2}$$

$$\therefore \quad \left(\frac{\partial P}{\partial T}\right)_V = \frac{R}{V-b}$$

Substituting, we get

$$\left(\frac{\partial E}{\partial V}\right)_T = \left[T\left(\frac{\partial P}{\partial T}\right)_V - P\right] \qquad \ldots(29)$$

$$= \left[\frac{TR}{V-b} - P\right]$$

On substitution, we get

$$\left(\frac{\partial E}{\partial V}\right)_T = \left[\frac{RT}{V-b} - \frac{RT}{V-b} + \frac{a}{V^2}\right]$$

$$= \frac{a}{V^2} \qquad \ldots(30)$$

It means that for actual gases the internal energy increases with increase in volume at constant temperature.

6.4 Partial Molar Quantities of an Ideal Gas in a Mixture of Ideal Gases

Expressions can be derived for various partial molar quantities of any component *'i'* in a mixture of ideal gases.

Starting from the equation

$$\mu_{i(\text{mix})(T.P.)} = \mu_{i^\circ(T)} + RT\ln P + RT\ln x_i \qquad \ldots(1)$$

Differentiating (1) w.r.t. temperature keeping P and composition constant,

$$\left[\frac{\partial \mu_{i(mix)(T.P.)}}{\partial T}\right]_{P,n_1,n_2\ldots} = \left[\frac{\partial \mu_{i^\circ(T)}}{\partial T}\right]_P + R\ln + P + R\ln x_i$$

or $$\overline{S}_t = -S_i^{\circ} + R\ln P + R\ln x_i$$

or $$\overline{S}_{i(mix)} = \overline{S}_{i(pure)} - R\ln x_i$$

where $$S_{i(pure)} = S_{i^\circ} - R\ln P$$

If we differentiate (i) w.r.t. P at constant 'l' and composition, we get

$$\left[\frac{\partial\mu_{i(mix)(T.P.)}}{\partial T}\right]_{P,n_1,n_2...} = \frac{RT}{P}$$

or $$\overline{V}_{i(mix)} = \frac{RT}{P} = \overline{V}_{i(pure)}$$

Thus the partial molar volume of an ideal gas is the same in the mixture as the partial molar volume in pure state.

For n moles of an ideal gas, we have

$$PV = nRT$$

or $$V = \frac{nRT}{P}$$

So, we get $\overline{V}_i = \frac{V}{n}$

Where n is the total number of moles of all gases in the mixture of ideal gases.

6.5 Chemical Potential (μ)

It is the most useful partial molar property. As already discussed the partial molar free energy $\overline{G}_i$ is known as chemical potential and is represented by symbol μ for the component so,

$$\mu_i = \left(\frac{\partial G}{\partial n_i}\right)_{T,P,n_j} \quad j \neq i \qquad ...(1)$$

For a system of n components, we can write

$$dG = \left(\frac{\partial G}{\partial T}\right)_{Pn_j} dT + \left(\frac{\partial G}{\partial P}\right)_{Tn_j} dP + \sum_{\iota=1}^{\kappa}\left(\frac{\partial G}{\partial n_i}\right)_{T,P_{nj}} dn_i$$

$$j \neq i \qquad ...(2)$$

or $$dG = \left(\frac{\partial G}{\partial T}\right)_{Pn_j} dT + \left(\frac{\partial G}{\partial P}\right)_{Tn_j} dt + \sum_{\iota=1}^{\kappa}\mu_i dn_i \qquad ...(3)$$

For a closed system dn_1, dn_2, dn_3...dn_j are zero. Hence equation (3) becomes

$$dG = \left(\frac{\partial G}{\partial T}\right)_{T,n_1n_2} dT + \left(\frac{\partial G}{\partial P}\right)_{T,n_1,n_2} dP \quad ...(4)$$

From our previous knowledge we have

$$dG = VdP\text{-}SdT \quad ...(5)$$

So from (4) and (5), we get

$$\left(\frac{\partial G}{\partial T}\right)_{P,n_1n_2} = -S \quad ...(6)$$

and

$$\left(\frac{\partial G}{\partial P}\right)_{T,n_1n_2} = V \quad ...(7)$$

From equation (3), (6) and (7), we get

$$dG = -SdT + VdP + \mu_1 dn_1 + \mu_2 dn_2 + ...$$
$$= -SdT + VdP + \Sigma\mu_i dn_i$$

Chemical Potential of an Ideal Gas

The chemical potential μ of an ideal gas at a given temperature is a function of its pressure P expressed in atmosphere. It can be expressed as

$$\mu_{(T,P)} = \mu^{o}{}_{(T)} + RT \ln P \quad ...(1)$$

$\mu_{(T)}{}^{o}$ is the standard chemical potential of the gas at 1 atmosphere pressure. Thus if the pressure in two regions of a gas phase in different there will be a spontaneous flow of gas from the region of higher pressure to a region of lower pressure till the pressure everywhere is same throughout the gaseous system.

The above relation (1) can also be derived in the following manner

$$V = (n_1 + n_2 + ...)\frac{RT}{P}$$

Differentiating w.r.t. n_i, we get

$$\left(\frac{\partial V}{\partial n_i}\right)_{T,P,n_1,n_2...} = \frac{RT}{P}$$

or

$$\overline{V}_i = \frac{RT}{P}$$

We have already calculated

$$\overline{V}_i = \left(\frac{\partial \mu_i}{\partial P}\right)_{T,n_1,n_2} \quad \text{[see eq. (12)]}$$

$$\left(\frac{\partial \mu_i}{\partial P}\right)_{T,n_1,n_2} = \frac{RT}{P}$$

Integrating this we get

$$\mu_i = \mu_i^{\circ} + RT \ln P$$

μ_i° is the integration constant and is equal to the standard chemical potential of the gas.

Chemical Potential of an Ideal Gas in a Mixture of Ideal Gases

Such an expression can be easily obtained if the pressure P in Eq. (1) replaced by p_i the partial pressure of the gas in the gaseous mixture. Thus the required equation is

$$\mu_{(mix)(T,P)} = \mu^{\circ}_{i(T)} + RT \ln p_i$$

but $$p_i = x_i \times P$$

$\therefore$ $$\mu_{(mix)(T,P)} = \mu_i{}^{\circ}{}_{(T)}\ RT \ln P + RT \ln x_i$$

or $$\mu i_{(mix)(T,P)} = \mu i_{pure(T,P)} + RT \ln x_i$$

$$\left[\because\ \mu i_{pure(T,P)} = \mu_i^{\circ} + RT \ln P\right]$$

Since x_i always less than 1 so $RT \ln x_i$ is always negative and hence

$$\mu i_{(mix)(T,P)} < \mu i_{(pure)(T,P)}$$

So in a mixture of ideal gases the chemical potential of an ideal gas is less than its chemical potential in pure state, provided the total pressure of the gaseous mixture and the pressure of the pure ideal gas are same.

It can be concluded that if a pure ideal gas and a mixture of ideal gases are separated from each other with the help of a permeable membrane which allows the pure to flow through it then the pure gas will flow to the mixture of the ideal gases. This also explains the spontaneous mixing of gases.

7

Activity, Activity Coefficient and Fugacity

7.1 ACTIVITY AND ACTIVITY COEFFICIENTS

Activity of a Gas. The *fugacity 'f'* and the free energy per mole of a gas (G) are related to each other according to the following equation,

$$G = RT \ln f + B \quad \ldots(1)$$

where B is a constant and it depends upon the temperature and the nature of the substance.

Since we can not evaluate the absolute value of 'G' so it is not possible to evaluate B. However, if all the measurements of free energy are made with reference to standard state, the energy so evaluated is know as G°. Let G° and f° be the standard free energy and standard state fugacity of the substance. So it becomes

$$G^{o} = RT \ln f^{o} + B \quad \ldots(2)$$

From (1) and (2), we get

$$G - G = RT \ln \frac{f}{f^{o}}$$

or

$$G = G + RT \ln \frac{f}{f^{o}}$$

$$= G + RT \ln a \quad \ldots(3)$$

where 'a' is called the activity

Thus $$'a' = \frac{f}{f^{o}}$$

Activity is thus defined as the ratio of fugacity of a substance in a particular state to the fugacity of the same substance in a standard state,

In the standard state,

$$G = G^{o}$$

In such a state equation (3) becomes

$$G^{o} = G + RT \ln a$$

Thus $$RT \ln a = 0$$

or $$a = 0$$

We thus conclude that the *standard state of a substance is that state when its activity (a) is equal to unity.*

Consider a mole of a substance having G_1 free energy in the initial state and G_2 be its free energy in the final state. Thus free change is given by

$$G_2 - G_1 = \Delta G = (G^{o} + RT \ln a_2) - (G^{o} + RT \ln a_1)$$

where a_1 and a_2 are the activities of the substance in initial and final states.

The above equation on simplification becomes

$$\Delta G = RT \ln \frac{a_2}{a_1} \quad \text{...(4)}$$

The change in free energy was calculated

as $$\Delta G = RT \ln \frac{p_2}{p_1} \quad \text{...(5)}$$

From equations (4) and (5) it is evident that pressure terms occurring in the equation for ideal gases have been replaced by activity terms in case of real gases.

So activity of a gas is a thermodynamic counterpart of pressure of a gas.

Activity Coefficient of a Gas

For ideal gas $a = P$ but real gases $A \propto P$

or $$a = \gamma P$$

γ is known as activity coefficient,

Relation between fugacity and activity Coefficient of a gas.

We have seen earlier that

$$G = G^{o} + RT \ln \frac{f}{f^{o}} \qquad \text{[equation 2]}$$

In the standard state of a gas $G^{o} = 1$ thus the above equation becomes

$$G = G^{o} + RT \ln f \qquad [\because f^{o} = 1]$$

So in case of an ideal gas fugacity and pressure are equal.

At very low pressure all gases approach an ideal gas behaviour

Thus $f = P$ when $P \rightarrow 0$

i.e., $$\underset{P \rightarrow 0}{Lt} \frac{f}{P} = 1$$

This ratio $\frac{f}{P}$ is called activity coefficient (γ) of a gas

So $$\gamma = \frac{f}{P}$$

At very low pressure in case of real gases or in case of ideal gases $\gamma = 1$.

More the deviation of γ of a gas from unity more will be its non-ideality.

Choice of Standard State

The activity of a substance is always related to its activity in the standard state. If different standard states are chosen we will get different values for activity. Thus the nature of standard state has a thermodynamic importance. For two phases we select the standard states as under:

(i) For a gas the standard state at a constant given temperature is one when the fugacity (f) of the gas is equal to one.

or $$a = \frac{f}{f^{o}}$$

$\therefore$ if $f^{\circ} = 1$ then $a = f$

so the activity of a gas is equal to its fugacity in the standard state.

For any gaseous component i, in the standard state $f_i = a_i = 1$.

Variation of Activity and Activity Coefficient of a Gas

(a) *Variation of activity of a Gas with Pressure*

We know

$$\mu_i = \mu_i^{\circ} + RT \ln a_i$$

Differentiating with respect to pressure keeping temperature and composition (N) constant, we get

$$\left(\frac{\partial \mu_i}{RT}\right)_{T,N} = RT\left(\frac{\partial \ln a_i}{\partial P}\right)_{T,N}$$

But $$\left(\frac{\partial \mu_i}{RT}\right)_{T,N} = \overline{V}_i \qquad \text{(Definition)}$$

$$\therefore \quad \overline{V}_i = RT\left(\frac{\partial \ln a_i}{\partial P}\right)_{T,N}$$

$$\frac{\overline{V}_i}{RT} = \left(\frac{\partial \ln a_i}{\partial P}\right)_{T,N}$$

$\overline{V}_i$ is the partial molar volume of the gas (i) in the mixture. This equation gives a relationship between pressure and activity of a gas.

(b) *Variation of activity of a gas with temperature*

We know

$$\mu_i = \mu_i + RT \ln a_i$$

Differentiating w.r.t. temperature keeping P and N (composition) constant, we get

$$\left(\frac{\partial \mu_i}{RT}\right)_{P,N} = RT\left(\frac{\partial \ln a_i}{\partial P}\right)_{P,N}$$

But $\left(\frac{\partial \mu_i}{\partial T}\right)_{P,N} = \bar{S}_i$ (definition)

$\therefore \quad -\bar{S}_i = \left(\frac{\partial \ln a_i}{\partial P}\right)_{P,N}$

or $\left(\frac{\partial \ln a_i}{\partial P}\right)_{P,N} = -\frac{\bar{S}_i}{RT}$

where $\bar{S}_i$, is partial molar entropy of gas (i) in the mixture

so $$\ln a_i = \frac{\mu_i}{RT} - \frac{\mu_i^\circ}{RT}$$

Differentiating this equation w.r.t. temperature keeping P and N constant, we get

$$\left(\frac{\partial \ln a_i}{\partial P}\right)_{P,N} = \frac{1}{R}\left[\frac{\partial(\mu_i / T)}{\partial T}\right]_{P,N} - \frac{1}{R}\left[\frac{\partial(\mu_i^2 / T)}{\partial T}\right]_{P,N}$$

or $$\left(\frac{\partial \ln a_i}{\partial P}\right)_{P,N} = \frac{1}{R}\left(-\frac{\bar{H}_i}{T^2}\right) - \frac{1}{R}\left[\frac{-\bar{H}_{i(ideal)}}{T^2}\right]$$

where $$-\frac{\bar{H}_i}{T^2} = \left[\frac{\partial(\mu_i / T)}{\partial T}\right]_{P,N}$$

and $$-\frac{-\bar{H}_{i(Ideal)}}{T^2} = \left[\frac{\partial(\mu_i / T)}{\partial T}\right]_{P,N}$$

or $$\left(\frac{\partial \ln a_i}{\partial T}\right) = \frac{\bar{H}_{i(Ideal)} - \bar{H}_i}{RT^2}$$

These equations gives variation of activity with temperature.

(c) *Variation of activity coefficients of a gas with pressure*

We know $\gamma = \frac{a_i}{p_i}$

$$\ln \gamma = \ln a_i - \ln p_i$$

But $$p_i = n_i P$$

where N_i is the mole fraction of component (i) and P is the total pressure and p_i is the partial pressure of component '*i*' so from these two equations, we have

$$\ln \gamma = \ln a_i - \ln N_i P$$
$$= \ln a_i - \ln N_i - \ln P$$

differentiating w.r.t. P at constant temperature and composition (N)

$$\left(\frac{\partial \ln \gamma}{\partial P}\right)_{T,N} = \left(\frac{\partial \ln a_i}{\partial P}\right)_{T,N} - \left(\frac{\partial \ln P}{\partial P}\right)_{T,N}$$

Since N_i does not depend on pressure its differential is zero.

$$\therefore \left(\frac{\partial \ln \gamma}{\partial P}\right)_{T,N} = \frac{\overline{V}_i}{RT} - \frac{1}{P}$$
$$= \frac{\overline{V}_i}{RT} - \frac{V_i}{RT}$$
$$= \frac{\overline{V}_i - V_i}{RT}$$

This equation gives the variation of activity coefficient with pressure.

7.2 Fugacity (f)

If (*f*) can be defined as

$$dF = nRTd \ln f \qquad ...(1)$$

Integrating, we get

$$F = nRT \ln f + B \qquad ...(2)$$

where B is a constant of integration

When $f = 1$, $G = +G_0$ then equation (9) becomes

$$G_0 = B + O \text{ or } G_0 = B$$

$$\therefore \quad G = G_0 + nRT \ln f$$

where G_0 is the free energy on *n* moles of real gas when its fugacity happens to be unity.

Thus 'fugacity' is a 'fictitious pressure' which is used in order to retain for real gases simple form of equation which are applicable to ideal gases only.

Fugacity at Low Pressure

$$\text{Fugacity} \propto \text{Pressure}$$

$$f \propto P$$

This relation has already been derived for ideal gases. In case of real gases fugacity (f) is initially less than pressure (P) and becomes greater than P at higher pressure. This deviation from ideal gas behaviour can be correlated with the Vander Waals constants 'a' and 'b'. At low pressure the Vander Waals equation may be written as

$$\left(P+\frac{a}{V^2}\right)(V) = RT \text{ [forces of attraction are important]}$$

$$Z(\text{Compressibility factor}) = \frac{PV}{RT}$$

$$= 1-\frac{a}{VRT}$$

So $Z < 1$

$\therefore$ f is less than P because of constant 'a'.

At higher pressure, the Vander Walls equation may be written as

$$P(V-b) = RT \quad \text{(excluded volume is important)}$$

or

$$Z = \frac{PV}{RT} = 1+\frac{bP}{RT}$$

So fugacity is greater than P. It is due to constant 'b'.

So we find that neglecting 'b' in Vander Waals equation makes 'f' less than P and neglecting 'a' makes 'f' greater than P.

It can be shown easily that for a Vander Waals gas following relations hold good

$$\text{(i) } \ln f = \ln\frac{RT}{V-b}+\frac{b}{V-b}-\frac{2a}{RTV}$$

and (ii) $\ln f = \ln P - \frac{aP}{(RT)^2} + \frac{bP}{RT} + \frac{abP^2}{2(RT)^3}$

The *compressibility factor 'z'* used above can be defined as the ratio V and V' at the same T and P. Thus

$$Z = \frac{V}{V'}$$

But $$V' = \frac{RT}{P}$$

so $$Z = \frac{PV}{RT}$$

$$Z = 1 \text{ for ideal gas.}$$

Z is a state function and is defined in terms of T,P,V. Various equations for Z as function of P, T and composition (or V, T and composition or P, V and composition) constitute *thermal equation of state*. Now an effort will be made to develop equations for calculations of 'f' or ϕ for a pure component or for a constant composition mixture from *thermal equations of state.*

In this equations subscript 'l' has been deleted first for simplicity. However, these equations are valid for both cases.

We know

$$dG = RT\, d\,(\ln f) \text{ at constant } T. \quad \ldots(10)$$

Also for one mole of solution at constant temperature and composition, we know

$$dG(T,N) = VdP \quad \ldots(11)$$

From (10) and (11), we get

$$RTd\,(\ln f) = Vd\,P \quad \ldots(12)$$

We also know

$$dG = RTd\,(\ln f) \text{ at Const. } (T)$$

From equation (3), we get

$$d\,(\ln f) = d(\ln f) + d\,(\ln P)$$

or $$d(\ln f) = d(\ln f) + \frac{dP}{P} \quad \ldots(13)$$

substituting (13) in (12) and using definition of Z, we get

$$d(\ln \phi) = (Z-1)\frac{dP}{P} \text{ (const. } T, N) \quad ...(14)$$

Equation (14) can be integrated from zero pressure state [where $\phi = 1$] to a pressure P state when we get

$$\ln \phi = \int_0^P (Z-1)\frac{dP}{P} \text{ (constant T, N)} \quad ...(15)$$

Since $f = \phi P$

We get

$$\ln (f) = \ln P + \int_0^P (Z-1)\frac{dP}{P} \text{(const } T, N) \quad ...(16)$$

Similar equations for which V, rather than P is variable can be easily derived. These are

$$\ln\phi = (Z-1) - \ln Z \int_\alpha^V (Z-1)\frac{dV}{V} \text{(const, T,N)} \quad ...(17)$$

$$\text{and } \ln\phi = \ln\frac{RT}{V} + (Z-1) - \int_\alpha^V (Z-1)\frac{dV}{V} \text{ (const T,N)} \quad ...(18)$$

In case of ideal gases Z = 1

$\therefore$ $\phi_i = \phi = 1$

and $fi = f = P$

For ideal gas mixtures analogous equations are

$\hat{\psi}_i = 1$

and $\hat{fi} = x_i P$

where x_i is the mole fraction of component '*i*' in the gas phase.

7.3 Determination of Fugacity

Some of the methods used for the determination of fugacity of gases and liquids mixture are described in following pages.

1. *Graphical Method.* As already proved in case of real gases at very low pressure $f = p$

But $dG = VdP$

$\therefore$ $RTd \ln f = VdP$ [for one mole]

or $$\left(\frac{\partial \ln f}{\partial P}\right)_T = \frac{V}{RT} \qquad \text{...(19)}$$

Introducing α where $\alpha = \dfrac{RT}{P} - V$, we get

$$RTd\ln f = \left(\frac{RT}{P} - \alpha\right)dP\left[\because V = \frac{RT}{P} - \alpha\right] \qquad \text{...(20)}$$

From (19) and (20), we get

$$RTd\ln f = \left(\frac{RT}{P} - \alpha\right)dP \qquad \text{...(21)}$$

Integrating (21) we get

$$RT\int d\ln f = RT\int_0^P \frac{dP}{P} - \int_0^P \alpha\, dP$$

or $$\int d\ln f = \int_0^P \frac{dP}{P} - \frac{1}{RT}\int_0^P \alpha\, dP$$

$$\ln f = \ln P - \frac{1}{RT}\int_0^P \alpha\, dP$$

To evaluate the integral a graph is plotted between a/RT and P as shown in Fig. 7.1 The area under the curve gives the values of integral and thus '*f*' can be calculated.

2. *Approximate Calculation Method.* At moderately low pressure, we can derive from Vander Wall's equation that PV for a real gas in a linear function of its pressure it temperatures is kept constant.

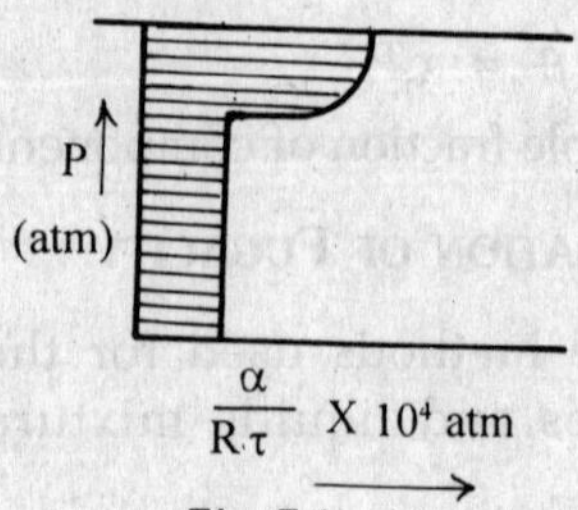

Fig. 7.1

Mathematically, we can write

$$PV = RT - \alpha P \qquad \rightarrow \left(\frac{\alpha}{R\tau} \times 10^4\right) \text{atm}^{-1}$$

or
$$\alpha = \frac{RT}{P} - V$$

a is constant over this moderate pressure range.

Applying this condition to Eq. (22), we get

$$\ln f = \ln P - \frac{\alpha}{RT}\int_0^P dP \qquad [\because \alpha \text{ is constant}]$$

or
$$\ln = \ln P - \frac{\alpha}{RT}P$$

or
$$\ln\frac{f}{P} = -\frac{\alpha P}{RT} \qquad \ldots(23)$$

At moderate pressure $\frac{f}{P} = 1$

Hence $\ln\frac{f}{P} = \frac{f}{P} - l \qquad [\log x \approx (x-1) \text{ when } x \to 1]$

From equation (23), we have

$$\frac{f}{P} - l = \frac{aP}{RT} \qquad \ldots(24)$$

Substituting in Eq. 24 value of α from the mathematical from given above, we get

$$f = -\frac{P^2V}{RT} \qquad \ldots(25)$$

From this equation fugacity of a gas can be calculated approximately if its pressure and molar volume are known. These results are quite accurate at low pressure and high temperature.

(3) *Generalized Method.* We have derived the equation,

$$a = \frac{RT}{P} - V = \frac{RT}{P}\left(1 - \frac{VP}{RT}\right)$$

The compressibility Z has been defined as $Z = \frac{PV}{RT}$

Form the above two equations, we get

$$a = \frac{RT}{P}(1 - Z) \qquad \ldots(26)$$

Putting this value of α in equation

$$\ln f = \ln P - \frac{1}{RT}\int_0^P \alpha \, dP \qquad \text{[Eq. 22]}$$

We get

$$\ln f = \ln P - \int_0^P \frac{1-Z}{P} dP \qquad \ldots(27)$$

Replacing pressure by reduced pressure, we get

$$\ln f = \ln P + \int_0^\pi \frac{Z-1}{\pi} d\pi \qquad \left[\pi = \frac{P}{P_c}\right]$$

or

$$\ln \frac{f}{P} = \int_0^\pi \frac{Z-1}{\pi} d\pi \qquad \ldots(28)$$

We know that the compressibility factors of all gases are approximately equal at given reduced temperature and pressure. Thus a graph can be lotted by keeping $\frac{f}{P}$ as ordinate and π as abscissa, θ as the parameter [Fig. 7.2]. To draw Fig. 7.2 the values of Z are derived from Fig. 7.3 and the Integral in equation (28) is evaluated.

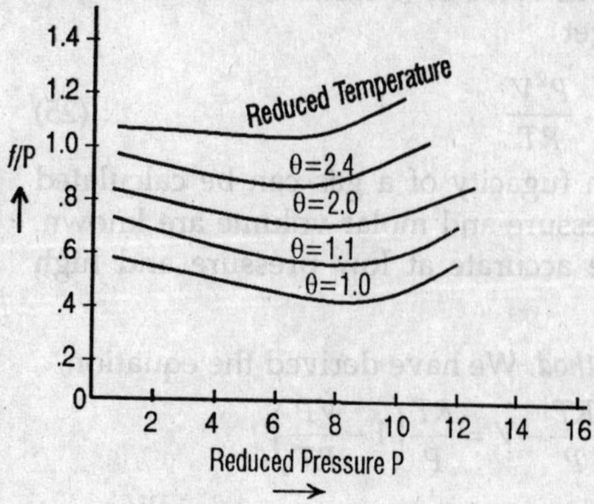

Fig. 7.2 *Fig. 7.3*

(4) From equation of State. If *f and* f^o are the fugacities of a gas at pressure *P* and at low pressure P_0 respectively. Then from equation (9), we get

$$RT\int_{f_0}^{f} d\ln f = \int_{P_0}^{P} VdP$$

or $$\ln\frac{f}{f_0} = \frac{1}{RT}\int_{P_0}^{P} -VdP \quad \text{...(29)}$$

Now $$\int_{P_0}^{P} VdV = [PV]_{V_0}^{V}\int_{V_0}^{V} PdV \quad \text{[Integration by parts]}$$

$$PV - P_0V_0 - \int_{V_0}^{V} PdV \quad \text{...(30)}$$

If V_0 is volume at P_0 then

$$P_0V_0 = RT$$

Thus

$$\int_{P_0}^{P} VdV = PV - RT - \int_{V_0}^{V} PdV \quad \text{...(31)}$$

or

$$\ln\frac{f}{F_0} = \frac{1}{RT}\left[PV - RT - \int_{V_0}^{V} PdV\right]$$

At very low pressure we have

$$\ln\frac{f}{f_0} = \ln\frac{f}{f_0}$$

∴ $$\ln\frac{f}{P_0} = \frac{1}{RT}\left[PV - RT - \int_{V_0}^{V} PdV\right]$$

or

$$\ln f = \ln P_0 + \frac{1}{RT}\left[PV - RT - \int_{V_0}^{V} PdV\right] \quad \text{...(32)}$$

The integral in this equation is calculated from equation of state like Vander Waal's equation i.e.,

$$\left(P + \frac{a}{V^2}\right)(V - b) = RT$$

or $$P = \frac{RT}{V-b} - \frac{a}{V^2}$$

Multiplying this equation by dV and integrating, we get

$$\int_{V_0}^{V} PdV = \int_{V_0}^{V} \frac{RT}{V-b} dV - \int_{V_0}^{V} \frac{a}{V^2} dV$$

$$= RT \ln \frac{V-b}{V_0 - b} + \frac{a}{V} - \frac{a}{V_0} \qquad \text{...(33)}$$

Substituting from (33) in (32) we get

$$\ln f = \ln P_0 + \frac{1}{RT}\left[PV - RT - RT\ln\frac{V-b}{V_0} - \frac{a}{V}\right] \qquad \text{...(34)}$$

Again from the Vander Waal's equation, we have

$$PV - RT = \left[\frac{RTV}{V-b} - \frac{a}{V}\right] - RT$$

or

$$PV–RT = \left[\frac{RTV}{V-b} - RT\right] - \frac{a}{V}$$

$$= \left[\frac{RTb}{V-b} - \frac{a}{V}\right] \qquad \text{...(35)}$$

Substituting from Eq. 35 in Eq. 34

$$\ln f = \ln P_0 + \frac{1}{RT}\left[\frac{RTb}{V-b} - \frac{a}{V} - RT\ln\frac{V-b}{V_0} - \frac{a}{V}\right]$$

$$= \ln P_0 + \frac{b}{V-b} - \frac{2a}{RTV} - \ln\frac{V-b}{V_0}$$

$$= \ln P_0 + \frac{b}{V-b} - \frac{2a}{RTV} - [\ln(V-b) - \ln V_0]$$

$$= \ln P_0 + \frac{b}{V-b} - \frac{2a}{RTV} - \left[\ln(V-b) - \ln\frac{RT}{P_0}\right]$$

$$[\because P_0 V_0 = RT]$$

$$= \ln P_0 + \frac{b}{V-b} - \frac{2a}{RTV} - [\ln(V-b) - \ln RT - \ln P_0]$$

$$= \frac{b}{V-b} - \frac{2a}{RTV} + \ln\frac{RT}{V-b}$$

Fugacity can be calculated from this equation if the values of *'a'* and *'b'* are known.

Change of Fugacity with Temperature

If G and G_0 are the molar free energy of gas in a given state and in the standard state respectively. The standard state here is the vapour state at low pressure. The change in free energy when one mole of this substance is converted will be

$$G_0 - G = RT \ln \frac{f_0}{f} \quad \text{...(36)}$$

Differentiating this equation w.r.t T at constant P, we get

$$\left(\frac{\partial G_0}{\partial T}\right)_P - \left(\frac{\partial G}{\partial T}\right)_P = R \ln \frac{f_0}{f} + RT\left(\frac{\partial \ln f_0}{\partial T}\right)_P - TR\left(\frac{\partial \ln f}{\partial T}\right)_P \quad \text{...(37)}$$

At very low pressure f_0 becomes independent of temperature

$$\therefore \quad \left(\frac{\partial \ln f_0}{\partial T}\right)_P = 0$$

From equation (36) we get

$$R \ln \frac{f_0}{f} = \frac{G - G_0}{T} \quad \text{...(38)}$$

Substituting from (38) in (37), we get

$$\left(\frac{\partial G_0}{\partial T}\right)_P - \left(\frac{\partial G}{\partial T}\right)_P = R \ln \frac{f_0}{f} - RT\left(\frac{\partial \ln f}{\partial T}\right)_P \quad \text{...(39)}$$

or

$$R \ln \frac{f_0}{f} = \frac{G_0 - G}{T} \quad \text{...(40)}$$

From equation (39), and (40)

$$\left(\frac{\partial G_0}{\partial T}\right)_P - \left(\frac{\partial G}{\partial T}\right)_P = \frac{G_0 - H}{T} - RT\left(\frac{\partial}{\partial T} \ln f\right)_P$$

But $$\left(\frac{\partial G}{\partial T}\right)_P = \frac{G-H}{T}$$

$$\therefore \frac{G_0-H_0}{T}-\frac{G-H}{T} = \frac{G_0-G}{T}-RT\left(\frac{\partial}{\partial T}\ln f\right)_P$$

or $$\left(\frac{\partial}{\partial T}\ln f\right)_P = \frac{H_0-H}{RT^2} \quad \text{...(41)}$$

where $H_0 - H$ gives the ideal heat of vapourisation.

Change of Fugacity with Pressure

We know

$$G = RT\ln f + C$$

where C is a temperature dependent constant.

Differentiating above equation w.r.t. pressure at constant temperature

$$\left(\frac{\partial G}{\partial P}\right)_T = RT\left[\frac{\partial \ln f}{\partial P}\right]_T$$

or $$V = RT\left[\frac{\partial \ln f}{\partial P}\right]_T \qquad \left[V=\left(\frac{\partial G}{\partial P}\right)_T\right]$$

or $$\left[\frac{\partial}{\partial P}\ln f\right]_T = \frac{\overline{V}}{RT} \quad \text{...(42)}$$

Integrating this equation, we get

$$RT\ln\frac{f_2}{f_1} = \int_{P_1}^{P_2} VdP \quad \text{...(43)}$$

Thus fugacity at any other pressure can be calculated if fugacity at some pressure is know.

Determination of Fugacity of Gas in a Mixture

Lewis-Rendull Rule. The same principle as used for determination of fugacity of pure gas is involved in this determination.

We have already derived

$$\left[\frac{\partial}{\partial T}\ln f\right]_T = \frac{\overline{V}}{RT} \qquad \text{[equation (42)]}$$

$$\text{or} \quad RTd\ln f_i = \overline{V}_i dP \qquad ...(44)$$

where $\overline{V}_i$ is the partial molar volume of component '*i*' the gaseous mixture

$$\text{Then} \quad \alpha_i = \frac{RT}{P} - \overline{V}_i \qquad [\text{definition of } \alpha]$$

where P is the total pressure.

Eliminating $\overline{V}_i$ from the above two equations

$$d\ln f_i d\ln P = \frac{\alpha_i}{RT}dP \qquad ...(45)$$

If $f_i\, f_0$ are the fugacities of component (*i*), at P and P_o where P_o is very low pressure, Integrating the equation (45) at constant temperature and composition, we get

$$\ln\frac{f_i}{f_0} = \ln\frac{P}{P_0} - \frac{1}{RT}\int_{P_0}^{P}\alpha_i dP \qquad ...(46)$$

$$\text{But} \quad (P_0)_i = N_i P_0$$

$$\text{or} \quad \frac{P}{(P_0)_i} = \frac{P}{N_i P_0}$$

where N_i is the mole fraction of (i) in the mixture

Taking log of both sides, we get

$$\ln\frac{P}{(P_0)_i} = \ln\frac{1}{N_i} + \ln\frac{P}{P_0}$$

$$\text{or} \quad \ln\frac{P}{(P_0)_i} = \ln N_i + \ln\frac{P}{(P_0)_i}$$

$$\text{or} \quad \ln\frac{P}{(P_0)} = \ln N_i + \ln\frac{P}{(P_0)_i}$$

Thus we get

$$\ln\frac{f_i}{f_0} = \ln N_i$$

$$= \ln \frac{P}{(P_0)_i} \frac{1}{RT} \int_{P_0}^{P} \alpha_i dP$$

or $$\ln f_i - \ln f_0 = \ln N_i + \ln P - \ln (P_0)_i \frac{1}{RT} \int_{P_0}^{P} \alpha_i dP$$

or $$\ln f_i = \ln N_i + \ln P - \ln f_0 - \ln (P_0)_i - \frac{1}{RT} \int_{P_0}^{P} \alpha_i dP$$

or $$= \ln N_i + \ln P + \ln \frac{f_0}{(P_0)_i} - \frac{1}{RT} \int_{P_0}^{P} \alpha_i dP \quad ...(47)$$

At very low pressure

$$\frac{f_0}{(P_0)_i} = 1$$

$$\therefore \quad \ln \frac{f_0}{(P_0)_i} = 0$$

So equation (47) becomes

$$\ln f_i = \ln N_i + \ln P - \frac{1}{RT} \int_{P_0}^{P} \alpha_i dP \quad ...(48)$$

If $\overline{V}_i$ and P are known then the values of α_i at various pressures can be calculated and the integral in equation (48) can be evaluated graphically. Thus fugacity f_i of any gaseous component can be determine in a mixture.

If the gases are mixed at constant temperature, $\Delta V = 0$. Then V_i of each gas is equal to the molar volume. In such a case (48) becomes

$$\ln f_i' = \ln P \frac{1}{RT} \int_{P_0}^{P} \alpha_i dP$$

or $$\ln f_i = \ln N_i + \ln f'$$

or $$\ln f_i = \ln N_i f_i'$$

or $$f_i = N_i f_i'$$

In such a mixture the fugacity f_i of a component is equal to the products of its (N_i) mole fraction and its fugacity is pure state (f_i) at the total pressure of the mixture.

This statement is known as Lewis-Randell rule.

8

Specific Heat of Gases

8.1 Introduction

Specific heat of a substance is defined as *the amount of heat required to raise the temperature of 1g of the substance by 1°C.*

8.2 Specific Heat of Gases

The specific heat of a substance depends on the experimental conditions. In case of solids and liquids practically there is no change in volume on heating and thus the work of expansion can be neglected. But the case of gas an appreciable expansion takes place on heating and so there is an appreciable amount of work done as work of expansion (*PdV*) so it becomes quite important to tell, in case of gases, whether the specific heat refers to the heating at constant volume or at constant pressure. Depending an experimental conditions there are two values of specific heats of gases.

(i) Specific heat of gas at constant volume (C_v)

It is the amount of heat required to raise the temperature of one gram of a gas by 1°C at constant volume.

(ii) Specific heat at constant pressure (C_p)

It is defined as the amount of heat required to raise the temperature of one gram of a gas by 1°C at constant pressure.

These two differ from each other by the amount of heat used up in doing the work of expansion in second case.

So $C_p = C_v$ + work of expansion

The specific heat at constant pressure (C_p) is always greater than the specific heat at constant volume (C_v) by an amount equal to the thermal equivalent of the external work done by the gas. Thus

$$C_p - C_v = Q = W$$

where Q is the amount of heat used to do extra work and W is the external work done in the same units as Q. Difference between two specific heats is given by

$$C_p - C_v = R$$

where C_p, C_v and R are all expressed in the same units.

Example 1. *Find the value of gas constant R for one gram molecule of a gas.*

Solution. We know

1 mole of any gas = 22400 ml of gas at N.T.P

For one mole of gas the gas equation is

$$PV = RT$$

or

$$R = \frac{PV}{T}$$

$$= \frac{76 \times 13.6 \times 981 \times 22400}{273}$$

$$= 8.32 \times 10^7$$

$$= 8.31 \text{ Joules/mole } ^\circ\text{K}.$$

Variation of Specific Heat of Diatomic Gases with Temperature

It has been observed experimentally that the specific heat of diatomic gases like H_2, N_2, Cl_2 and HCl etc. increases with increase of temperature and decreases with decrease of temperature finally approaching the limiting value of $\frac{3}{2}$ (= 2.98 calories) which is the value of specific heat of monoatomic gases at all temperatures. Thus it leads us to conclude that at very low temperatures all diatomic gases have same value of specific heat of constant volume i.e., as that of monoatomic gas. Euken found experimentally that the specific heat of hydrogen at constant volume approaches $\frac{3}{2}$ R at about 60°K.

Example 1. *Assuming the ideal gas molecules as rigid, but capable of free rotation about either of two mutually perpendicular axis which are at right angles to the line joining the two atoms of the molecule. Find C_v, C_p and γ and compare γ with the experimental value of γ for a diatomic gas.*

Solution. If we assume that the molecules of a diatomic gas are rigid, then it can have only five degrees of freedom (three of translational and two rotational).

The kinetic energy per degree of freedom according to equipartitional principle of energy, is equal to $\frac{1}{2}kT$.

Therefore, kinetic energy associated with five degrees of freedom = $\frac{5}{2}kT$.

Thus K.E. per gram molecule = $\frac{5}{2}kT$.

and the K.E. per gram molecule = $\frac{5}{2}$N.k.T.

or $$W = \frac{5}{2}RT$$

Hence

$$\text{Specific heat } C_v = \frac{dW}{dT} = \frac{5}{2}R$$

$$C_p = C_v = R$$

$$= \frac{5}{2}R + R = \frac{7}{2}R$$

and $$\gamma = \frac{C_p}{C_v} = \frac{\frac{7}{2}R}{\frac{5}{2}R}$$

$$= \frac{7}{5}$$

$$= 1.40$$

The experimentally observed value of g at 15°C are given in the following table:

Gas	*Value of* γ
H_2	1.41
N_2	1.404
O_2	1.401
HCl	1.41
CO	1.404

These values of γ agree with those calculated theoretically.